Cyber Space and Outer Space Security

RIVER PUBLISHERS SERIES IN DIGITAL SECURITY AND FORENSICS

The "River Publishers Series in Security and Digital Forensics" is a series of comprehensive academic and professional books which focus on the theory and applications of Cyber Security, including Data Security, Mobile and Network Security, Cryptography and Digital Forensics. Topics in Prevention and Threat Management are also included in the scope of the book series, as are general business Standards in this domain.

Books published in the series include research monographs, edited volumes, handbooks and textbooks. The books provide professionals, researchers, educators, and advanced students in the field with an invaluable insight into the latest research and developments.

Topics covered in the series include:-

- Blockchain for secure transactions
- Cryptography
- Cyber Security
- Data and App Security
- Digital Forensics
- Hardware Security
- IoT Security
- Mobile Security
- Network Security
- Privacy
- Software Security
- Standardization
- Threat Management

For a list of other books in this series, visit www.riverpublishers.com

Cyber Space and Outer Space Security

Editors

Karthikeyan Periyasami
RV University, Bengaluru, India

Polinpapilinho F. Katina
University of South Carolina Upstate, USA

Rajagopal Ramasamy
Alliance University, India

River Publishers

Routledge
Taylor & Francis Group
NEW YORK AND LONDON

Published 2024 by River Publishers
River Publishers
Alsbjergvej 10, 9260 Gistrup, Denmark
www.riverpublishers.com

Distributed exclusively by Routledge
605 Third Avenue, New York, NY 10017, USA
4 Park Square, Milton Park, Abingdon, Oxon OX14 4RN

Cyber Space and Outer Space Security / Karthikeyan Periyasami, Polinpapilinho F. Katina and Rajagopal Ramasamy.

Routledge is an imprint of the Taylor & Francis Group, an informa business

ISBN 978-87-7004-188-1 (hardback)
ISBN 978-87-7004-645-9 (paperback)
ISBN 978-87-7004-637-4 (online)
ISBN 978-8-770-04628-2 (ebook master)

Contents

Preface

In the rapidly evolving landscape of modern technology, the convergence of cyberspace and outer space has introduced unprecedented challenges and opportunities. This book, "Cyber Space and Outer Space Security and Mitigation Strategies," addresses these dual realms, providing comprehensive insights into their similarities, differences, and the security imperatives they present.

Cyberspace, the virtual domain of interconnected computer systems and networks, and outer space, the physical expanse beyond Earth's atmosphere where satellites and spacecraft operate, are both pivotal to contemporary life and progress. As these domains become increasingly intertwined, their respective security concerns demand holistic and integrated strategies. Cyberspace security focuses on protecting digital information and communication systems, whereas outer space security ensures the physical safety of assets operating in the harsh environment of space. Despite their distinct natures, both areas are vital and interdependent, necessitating a coordinated approach to security.

The importance of this book stems from the recognition that both cyberspace and outer space face unique and significant threats, including complexity, lack of standardization, limited resources, and the need for international cooperation. This compilation brings together expertise from diverse fields to explore these issues comprehensively, offering practical strategies and frameworks to enhance security across both domains.

Key themes in this book include:

Cyberspace and Outer Space Security: Exploring the foundational concepts, their intersections, and the necessity for integrated security measures.

Cybersecurity for Space Systems: Addressing the sophisticated threats to space infrastructure and the evolving nature of space security since the Outer Space Treaty.

Cybersecurity Kill Chain in Outer Space: Adapting the Cybersecurity Kill Chain framework to the unique challenges of protecting space-based systems.

Ransomware Attacks and Mitigation: Examining the rising threat of ransomware in cyberspace and effective mitigation strategies.

Cyber Attacker Profiling and Cyberbullying: Profiling cyber attackers to anticipate and prevent attacks, and addressing the pervasive issue of cyberbullying.

Blockchain-enabled Drone Delivery in Space: Proposing innovative solutions for medical supply delivery in space using blockchain and drone technology.

Resilient Schemes for Autonomous Vehicles: Enhancing security against external attacks in autonomous vehicular networks.

Internet of Drones Security and Privacy: Securing the burgeoning field of drone technology integrated with IoT.

Post-quantum Cryptographic Approaches: Preparing for the future with cryptographic methods resilient to quantum computing threats.

This book is a product of collaboration among leading experts in cybersecurity, space security, and related fields. We extend our gratitude to all contributors for their invaluable insights and dedication. Their collective expertise forms the backbone of this work, making it an essential resource for researchers, policymakers, and practitioners aiming to safeguard the integrity of cyberspace and outer space.

We also acknowledge the support of our publishers, River Publishers, whose commitment to disseminating knowledge has made this publication possible.

Karthikeyan Periyasami
Polinpapilinho F. Katina
Rajagopal Ramasamy

Editors,
"Cyber Space and Outer Space Security and
Mitigation Strategies" 2024

List of Contributors

Ahmad, Awais, *Information Systems Department, Imam Mohammed Bin Saud University, KSA*

Anita, E. A. Mary, *Department of CSE, CHRIST (Deemed to be University), India*

Daniel J., Alfred, *Department of Computer Science and Engineering, Karpagam Academy of Higher Education, India*

Gautam, Minakshi, *Presidency University, India*

Geetha, A., *Department of Computer Science and Engineering, Alliance University, India*

Hariharan, U., *Department of Computer Science and Engineering, Apex Institute of Technology, Chandigarh University, India*

Jayakanth, J. J., *Department of Computational Intelligence, School of Computing, College of Engineering and Technology, SRM Institute of Science and Technology, India*

Jebadurai, Immanuel Johnraja, *Department of Computer Science and Engineering, Karunya Institute of Technology and Sciences, India*

Jenefa, J., *Department of CSE, CHRIST (Deemed to be University), India*

Karthikeyan, P., *School of Computer Science and Engineering, RV University, India*

Karthikeyan, S., *Department of Computer Science and Engineering, Faculty of Engineering and Technology, Jain (Deemed-to-be University), India*

Kavitha, V., *Department of Computer Science and Engineering, Faculty of Engineering and Technology, Jain (Deemed-to-be University), India*

Keerthika, V., *Department of Computer Science and Engineering, Alliance University, India*

Kesavamoorthy, R., *Department of Computer Science and Engineering, CMR Institute of Technology, India*

Mahalakshmi, R., *Presidency University, India*

Pallavi, R., *Presidency University, India*

Paulraj, Getzi Jeba Leelipushpam, *Department of Computer Science and Engineering, Karunya Institute of Technology and Sciences, India*

Poornima, Galiveeti, *Presidency University, India*

Raj, D. M. Deepak, *Department of Computer Science and Engineering, Alliance University, India*

Rajasekar, Vani, *Kongu Engineering College, India*

Rajkumar, M. Newlin, *Department of Electrical and Electronic Engineering, Anna University Regional Campus, India*

Rajkumar, K., *Department of Data Science and Business Systems, College of Engineering and Technology, India*

Sambandam, Rakoth Kandan, *Department of CSE, CHRIST (Deemed to be University), India*

Sarveshwaran, Velliangiri, *Department of Computational Intelligence, School of Computing, College of Engineering and Technology, SRM Institute of Science and Technology, India*

Shafik, Wasswa, *Dig Connectivity Research Laboratory (DCRLab), Uganda*

Sophia, S., *Department of Computer Science and Engineering, Karunya Institute of Technology and Sciences, India*

Vetriveeran, Divya, *Department of CSE, CHRIST (Deemed to be University), India*

Vignesh C., Chandru, *School of Computer Science and Engineering, Vellore Institute of Technology, India*

Vinodha, D., *Department of CSE, CHRIST (Deemed to be University), India*

List of Figures

List of Tables

List of Abbreviations

ABM	Anti-ballistic missile
AD&C	Attitude determination and control
AEAD	Authenticated Encryption with Associated Data
AES	Advanced Encryption Standard
AI	Artificial intelligence
AIA	Aerospace Industries Association
APT	Advanced Persistent Threats
ASAT	Anti-satellite
ASAT	Anti-satellite
ATO	Air tasking order
ATT&CK	Adversarial Tactics, Techniques, and Common Knowledge
AV	Autonomous vehicle
AVNET	Autonomous vehicular ad hoc network
BEDDS	Blockchain-enabled drone delivery system
C&DH	Command and data handling
CAVE	Cyber Analysis Visualization Environment
CBM	Confidence-building measure
CCoE	Cybersecurity Centre of Excellence
CDER	Cyber Defense Engineering and Research Group
CFS	Cryptographically-Feasible Systems
CNN	Convolutional Neural Networks
COMSEC	Communications security
CTBTO	Preparatory Commission for the Comprehensive Nuclear-Test-Ban Treaty Organization
CVP	Closest vector problem
DDoS	Distributed denial of service
DDQN	Double deep Q-network
DeFi	Decentralized finance
DER	Distributed energy sources
DFaR	Data flooding against ransomware
DKIM	DomainKeys Identified Mail
DL	Deep learning

DMARC	Domain Message Authentication Reporting & Conformance
DoS	Denial-of-service
DRA	Disaster relief authority
DSA	Digital signature algorithm
DSN	Deep space network
EA	Electronic attack
EB	Entangle base
ECC	Elliptic curve cryptography
ECDH	Elliptic curve DH
ECDSA	Edwards-curve Digital Signature Algorithm
ECPA	Efficient conditional privacy-preservation scheme without pairings
EDR	Endpoint detection and response
EPDS	Electrical power and distribution subsystem
EPP	Endpoint protection platform
ESA	European Space Agency
FANET	Flying Ad-Hoc Networks
FCSD	Fog computing aided swarm of drones
FKM	Fuzzy K means
FL	Federated learning
FPGA	Field-programmable gate array
FSW	Field switchable waveform
FS	Flight software
FTS	Few-time signature
GDPR	General Data Protection Regulation
GIS	Geographic information system
GNSS	Global navigation satellite systems
GPS	Global positioning system
HBS	Hash-based signature
HI	Human intelligence
HPC	Hardware performance counters
HS	Hierarchical signature
IBE	Identity-based encryption
IBOOS	ID-based online/offline signature
IBS	Identity-based signature
ICBM	Intercontinental ballistic missile
ICJ	International Court of Justice
ICS	Industrial control systems
IDS	Intrusion detection systems

IMU	Inertial Measurement Units
INFOSEC	Information security
IOC	Indicators of Compromise
IoD	Internet of Drones
IoT	Internet of Things
IP	Internet protocol
IPS	Intrusion prevention systems
IRP	I/O request packet
ISAC	Information Sharing and Analysis Center
ISR	Intelligence, surveillance, and reconnaissance
ISS	International Space Station
IT	Information technology
JPL	Jet Propulsion Laboratory
JWST	James Webb Space Telescope
KEM	Key encapsulation mechanism
LEO	Low Earth orbit
LSTM	Long Short-Term Memory
MANET	Mobile Ad-Hoc Networks
MFT	Master file table
MIT	Massachusetts Institute of Technology
ML	Machine learning
MLP	Multi-layer perception
MTS	Multi-time signature
MVD	Motor vehicles division
NASA	National Aeronautics and Space Administration
NATO	North Atlantic Treaty Organization
NIST	National Institute of Standards and Technology
OAOFS-MLIDS	Oppositional Aquila optimizer-based feature selection with machine learning enabled intrusion detection system
OBU	On-board unit
OCIO	Office of the Chief Information Officer
OPSEC	Operations security
OS	Operating system
OST	Outer Space Treaty
OTS	One-time signature
PAROS	Prevention of an Arms Race in Outer Space
PCSF	Power control in secure FL
PEDA	Predictive Early Detection and Analysis
PM	Prepare and measurement

PMS	Proactive monitoring system
PoW	Proof of Work
PQC	Post-Quantum Cryptography
PS	Propulsion subsystem
PUA	Potentially unwanted applications
PUP	Potentially unwanted program
QKD	Quantum key distribution
QoS	Quality of service
R2V	RSU-to-vehicle
RaaS	Ransomware-as-a-service
RDP	Remote desktop protocol
RSA	Rivest-Shamir-Adleman
RSU	Road side unit
SCM	Supply chain management
SIVP	Shortest independent vector problem
SKR	Secret key rate
SLS	Space Launch System
SMS	Structures and mechanisms subsystem
SP	Service provider
SPACF	Secure privacy-preserving authentication scheme with cuckoo filter
SPD5	Space Policy Directive
SPF	Sender policy framework
SSA	Social security administration
SSAW	Space situational awareness
SSD	Solid State Drives
SSID	Service set identifier
SVP	Shortest vector problem
SVP	Shortest vector problem
TA	Trusted authority
TCS	Thermal control subsystem
TRANSEC	Transmission security
TSMC	Taiwan Semiconductor Manufacturing Co.
TT&C	Telemetry, tracking, and command
TTPs	Tactics, techniques, and procedures
UAV	Unmanned aerial vehicle
UHBVN	Uttar Haryana Bijli Vitran Nigam
UNGA	United Nations General Assembly
UNOOSA	United Nations Office for Outer Space Affairs
USGS	United States Geological Survey

V2D	Vehicle to DRA
V2R	Vehicle-to-RSU
V2V	Vehicle-to-vehicle
VANET	Vehicular ad hoc network
WOTS	Winternitz One-Time Signature Scheme
XSS	Cross-site scripting

1

Cyberspace and Outer Space Security and Mitigation Strategies

R. Kesavamoorthy[1] and P. Karthikeyan[2]

[1]Department of Computer Science and Engineering, CMR Institute of Technology, India
[2]School of Computer Science and Engineering, RV University, India

Email: kesavamoorthycse@gmail.com; nrmkarthi@gmail.com

Abstract

This discussion covers the concepts of cyberspace and outer space security, including their similarities and differences. Cyberspace is the virtual realm of interconnected computer systems, networks, and the Internet, while outer space is the physical realm beyond Earth's atmosphere, where satellites, spacecraft, and other objects operate. Both depend on technology and are interconnected but have unique challenges and hazards. Cyberspace security primarily protects digital information and communication technology systems and information. In contrast, outer space security is focused on the physical security of the assets and ensuring outer space's safety. The limitations of cyberspace and outer space security, such as complexity, lack of standardization, international agreements, and limited resources, are discussed. The importance of the cooperation of the international entities and developing a framework to address security concerns in both areas is also highlighted.

1.1 Introduction

Cyberspace is the virtual environment created by computer networks and the Internet. It includes the Internet and the infrastructure that supports it, such as servers, routers, and data centers. It also includes the various systems and devices with the Internet, such as smartphones, laptops, and industrial control systems. Cyberspace also encompasses the information and data stored and transmitted on these networks, including personal information, financial data, and intellectual property. Outer space is the physical space beyond the Earth's atmosphere, where the Earth's gravity is no longer the dominant force. The vast expanse of space surrounds the Earth and the rest of the solar system. It is also known as "space" or "the universe" [1][2].

Cyberspace security protects internet-connected systems, including hardware, software, and data from theft, damage, or unauthorized access. This includes measures to prevent cyber-attacks, such as hacking and malware, and mitigating their effects. Outer space security protects spacecraft, satellites, and other assets in outer space from interference, damage, or destruction. This includes measures to prevent collisions and hostile actions, such as the jamming of signals or the destruction of satellites [3].

1.2 Cyberspace and Outer Space

Computer networks and the Internet create cyberspace. It is a complex and constantly evolving ecosystem that encompasses a wide range of activities and systems. Key components of cyberspace include:

- The Internet: It is the backbone of cyberspace. The Internet is the network of networks that enables the exchange of data. It is based on a set of protocols, such as TCP/IP, allowing different types of computer systems to communicate.
- Infrastructure: The physical components that make up cyberspace, including servers, routers, data centers, and other hardware that support the Internet. These systems are responsible for storing, processing, and transmitting data across the Internet.
- Systems and devices: The various systems and devices with the Internet, including smartphones, laptops, tablets, and industrial control systems. Individuals, businesses, and organizations use these systems and devices to access and share information and control and monitor a wide range of systems and processes.
- Data and information: The vast amount of information stored and transmitted on the Internet, including personal information, financial data, and intellectual property. Individuals, businesses, and organizations use this data and information.
- Cyber activities: The various activities in cyberspace, including communication, commerce, entertainment, and social interaction. This also

includes malicious activities such as hacking, cyber-attacks, and cyber espionage.

Overall, cyberspace is a vast and complex environment that is constantly evolving.

Outer space is the physical space beyond the Earth's atmosphere, where the Earth's gravity is no longer the dominant force. The vast expanse of space surrounds the Earth and the rest of the solar system. It is also known as "space" or "the universe."

Some of the key characteristics of outer space include:

- It is vast and mostly empty, with the nearest star about 4.2 light-years away from Earth.
- It is freezing, with temperatures approaching absolute zero.
- It is a vacuum, meaning no air or other gases are present.
- It is home to many celestial bodies, including planets, stars, galaxies, and other astronomical objects.
- It is also home to various artificial objects such as satellites, space debris, and space stations.
- It is also the environment where all space-related activities like space exploration, space mining, and space warfare occur.

Outer space is often studied by scientists and researchers to better understand the universe's origins and structure and to gain knowledge and resources that can be used to benefit humanity. In recent years, there has been a growing interest in the commercial use of outer space, with private companies and governments exploring opportunities in areas such as satellite communication, space tourism, and space mining [4][5][6].

1.2.1 Relation between cyberspace and outer space

Cyberspace and outer space are related in several ways:

- Both cyberspace and outer space are part of the global infrastructure that supports modern society. Cyberspace is the virtual environment created by computer networks, while outer space is the physical environment beyond the Earth's atmosphere. Both are essential to the functioning of modern society, and both are increasingly being used.
- Both cyberspace and outer space are used for communication. Cyberspace is used for communication through the Internet and other

computer networks, while outer space is used for satellite communication. Satellites in outer space are used for various communication services, including television and radio broadcasting, telephone and internet service, and global positioning systems.

- Both cyberspace and outer space are vulnerable to malicious activities. Using cyberspace and outer space for essential functions such as communication, navigation, and military activities makes them a target for malicious actors. Cyberspace is vulnerable to cyber-attacks, such as hacking, malware, and espionage. At the same time, outer space is vulnerable to physical attacks, such as the jamming of signals and the destruction of satellites.
- Both cyberspace and outer space are being used for research and exploration. Studying cyberspace and outer space leads to new understanding and knowledge of the world and the universe. Cyberspace is used to research new technologies and applications, while outer space is used to explore the solar system and beyond.

Overall, while cyberspace and outer space are distinct environments, they are closely related, and both are critical to the functioning of modern society. As technology advances, the relationship between cyberspace and outer space will continue to evolve and develop.

1.2.2 Difference between cyberspace and outer space

Cyberspace and outer space are distinct environments that have several key differences:

- Location: Cyberspace is a virtual environment created by computer networks, while outer space is the physical environment beyond the Earth's atmosphere.
- Nature: Cyberspace is made up of data, information, and technology that support the Internet and computer networks, while outer space is made up of physical celestial bodies and other astronomical objects.
- Accessibility: Cyberspace is easily accessible to people with internet access and a connected device. In contrast, outer space is only accessible through specialized equipment and technology such as spacecraft and satellites.
- Temperature: Cyberspace operates at room temperature, while outer space is frigid.

- Gravity: Cyberspace is not affected by gravity, while outer space is affected by the gravity of celestial bodies.
- Security concerns: Cyberspace is vulnerable to cyber-attacks, hacking, malware, and cyber espionage, while outer space is vulnerable to physical attacks, such as the jamming of signals, collisions, and destruction of satellites.
- International regulations: Cyberspace has international laws and regulations to govern its activities. However, outer space is governed by international treaties, such as the Outer Space Treaty, which regulates activities in outer space.

While cyberspace and outer space are critical to the functioning of modern society, they are distinct environments with different characteristics and challenges.

1.2.3 Cyberspace security

Cyberspace and outer space are related in several ways: cyberspace security includes measures to prevent cyber-attacks, such as hacking and malware, and mitigate their effects.

There are several critical elements of cyberspace security, including:

- Network security: Protecting the infrastructure that supports the Internet, including servers, routers, and data centers, from unauthorized access and attack.
- Device security: Protecting the various systems and devices with the Internet, such as smartphones, laptops, and industrial control systems, from unauthorized access and attack.
- Data security: Protecting the data and information stored and transmitted on the Internet, including personal information, financial data, and intellectual property, from unauthorized access and attack.
- Identity and access management: Controlling who has access to sensitive information and systems and ensuring that only authorized individuals can access them.
- Incident response: Having a plan for responding to a cyber-attack or other security incident and quickly and effectively responding to minimize the damage and recover from the attack.
- Compliance: Ensuring that organizations comply with the relevant laws, regulations, and industry standards regarding cybersecurity.

Overall, cyberspace security is a complex and constantly evolving field that requires a multi-disciplinary approach. It involves technology, policies, procedures, and human factors to ensure safety and security [7][8].

1.2.4 Outer space security

Outer space security protects outer space assets and activities from unauthorized interference, damage, or destruction. This includes the protection of spacecraft, satellites, and other space infrastructure.

There are several key elements of outer space security, including:

- Space awareness: Monitoring the space environment to detect and track objects in space, such as satellites, debris, and other space objects, to ensure safety and to prevent collisions.
- Space traffic management: Managing the safe and orderly flow of space traffic and minimizing the risks and other hazards to space assets and activities.
- Protection of space assets: Protecting spacecraft, satellites, and other space infrastructure from physical and cyber-attacks, including jamming of signals and spoofing of navigation systems.
- Space debris management: Addressing the issue of space debris, which poses a risk to space assets and activities, by promoting the responsible use of outer space and the reduction of space debris.
- Compliance with international laws and regulations: Ensuring that countries and organizations comply with international laws and regulations governing outer space activities.

Overall, outer space security is a complex and multidimensional issue involving the protection of assets and activities in outer space and the security of the space environment itself. It requires a combination of technical, legal, and diplomatic measures [9][10].

1.2.5 Mitigation strategies in cyberspace

Mitigation strategies in cyberspace refer to the set of measures that organizations and individuals can take to prevent, detect, and respond to cyber threats and attacks. They are designed to minimize the risk of a successful cyber-attack and the damage and disruption that can result from it [11][12][13].

Here are some examples of mitigation strategies in cyberspace:

- Risk management: This assesses the likelihood and impact of those risks and prioritizes them based on their potential impact. Once the risks have been identified, the organization can implement measures to reduce or eliminate them.
- Incident response planning: This includes roles and responsibilities for different incident response team members, as well as procedures for communicating with stakeholders and the media.
- Network security: Protecting the organization's network and systems is critical to cybersecurity. It also includes implementing secure configurations and policies to control sensitive systems and data access.
- Security awareness and training: Educating employees and other users about safe computing practices and the potential risks of cyber-attacks and training them to recognize and report suspicious activity.
- Patch management: Regularly update software and systems to address known vulnerabilities and prevent exploitation.
- Security monitoring: Continuously monitor the organization's systems and network for signs of suspicious activity and take appropriate action when an incident is detected.

1.2.6 Mitigation strategies in outer space

Mitigation strategies in outer space refer to the set of measures that organizations and countries can take to prevent, detect, and respond to threats and attacks in outer space. They are designed to minimize the risk of unauthorized interference, damage, or destruction to outer space assets and activities.

Here are some examples of mitigation strategies in outer space:

- Space situational awareness (SSA): SSA is the process of monitoring and tracking objects in the space environment to ensure spaceflight safety and prevent collisions. This includes maintaining a catalog of objects in space and providing information to space operators to help them avoid collisions and other hazards. SSA also includes detecting and tracking debris and other space objects and providing information to help mitigate the risk of collisions.
- Space traffic management: Space traffic management manages the safe and orderly flow of space traffic. This includes providing information

to space operators to help them avoid collisions and other hazards and coordinating and communicating with other space operators to ensure safe and efficient use of outer space.

- Protection of space assets: Protecting spacecraft, satellites, and other space infrastructure from physical and cyber-attacks is essential for outer space security. This includes implementing technical and operational measures to protect space assets from signal jamming, spoofing navigation systems, and other cyber-attacks. It also includes physical protection of space assets from collisions, debris, and other hazards.
- International cooperation: Promoting international cooperation is essential for outer space security. This includes sharing information and best practices with other countries and space organizations and developing confidence-building measures and transparency.
- Compliance with international laws and regulations: Ensuring that countries and organizations comply with international laws and regulations governing outer space activities.
- Space debris management: Implementing measures to minimize the creation of space debris and promote the responsible use of outer space.
- Prevention of arms race in outer space: Limiting the development, testing, and deployment of weapons in space and promoting transparency and confidence-building measures among states.

Overall, outer space security is a complex and multidimensional issue involving the protection of assets and activities in outer space and the security of the space environment itself. It requires a combination of technical, legal, and diplomatic measures to ensure the safety of outer space assets and activities and to prevent the weaponization of outer space [14][15][16][17].

1.2.7 Cybersecurity Kill Chain

The Cyber Kill Chain, also known as the Lockheed Martin Cyber Kill Chain, is a model that describes the stages of a typical cyber-attack. The model is intended to help organizations understand how cyber-attacks happen and to identify potential vulnerabilities and weaknesses in their defenses. The stages of the Cyber Kill Chain include:

1. Reconnaissance: The attacker researches and gathers information about the target organization and its systems and networks. This can include

using social media, public records, and other open-source information to identify potential targets and vulnerabilities.

2. Weaponization: The attacker develops and delivers the tools and methods necessary to exploit the vulnerabilities identified during reconnaissance. This can include creating malware, phishing emails, and other attack vectors.

3. Delivery: The attacker delivers the weaponized tools and methods to the target organization. This can include sending phishing emails, exploiting web browser or software vulnerabilities, and using other methods to access the target's systems.

4. Exploitation: The attacker uses the tools and methods delivered during delivery to access the target's systems and networks. This can include executing malware, stealing login credentials, or exploiting other vulnerabilities.

5. Installation: The attacker establishes a foothold in the target's systems and networks and moves laterally through the organization. This can include installing backdoors, creating new user accounts, or deploying additional malware.

6. Actions on objective: The attacker uses the compromised systems and networks to achieve their ultimate goal.

The Cyber Kill Chain is a model that describes the stages of a typical cyber-attack, and it is primarily used to understand and mitigate cybersecurity threats. While it is a well-established model in the field of cybersecurity, it has yet to be widely applied to outer space security. Outer space security is a relatively new field with unique challenges and hazards that are different from cybersecurity.

Outer space security concerns include threats such as space debris, space weather, interference of signals and jamming, and space situational awareness, among others. In outer space, the security concerns are more focused on the physical security of the assets and ensuring the safe and responsible use of outer space.

However, it is essential to note that as space technology and the integration of space systems with computer systems and networks continue to grow, the risk of cyber-attacks in outer space will also increase. In this case, the application of the Cyber Kill Chain model to outer space security could be helpful in understanding and mitigating the cyber threats to space systems and assets.

In summary, while the Cyber Kill Chain model is primarily used in cybersecurity and not explicitly designed for outer space security, it could be adapted and applied to better understand and mitigate the cyber threats to space systems and assets as the integration of space technology with computer systems and networks continues to grow [13][14].

1.2.8 Limitations of cyberspace security

Cyberspace security has several limitations that make it challenging to protect against cyber-attacks. Some of these limitations include:

1. Complexity: The constantly evolving nature of technology and interconnected systems make securing all possible vulnerabilities and attack vectors difficult.
2. Human error: People are often the weakest link in cybersecurity, and mistakes such as falling for phishing scams or not updating software can leave systems vulnerable to attack.
3. Lack of standardization: There needs to be more standardization in terms of security protocols, and the numerous different platforms, operating systems, and devices used in cyberspace make it challenging to implement comprehensive security measures.
4. Lack of international agreements: There needs to be more international agreements and laws to govern cyberspace, which makes it challenging to hold actors accountable for cyber-attacks or to cooperate in investigating and preventing them.
5. Lack of visibility: It can be difficult as attackers can use various techniques to hide their tracks and evade detection.
6. Limited resources: Smaller organizations, governments, and countries may need more resources and expertise to defend themselves from cyber-attacks.
7. Cyber-attacks are constantly evolving: Cybercriminals and hackers are constantly developing new techniques to evade detection and infiltrate systems, making it difficult for organizations to keep pace with new threats.
8. Dependence on technology: As the world becomes increasingly dependent on technology, the risk of cyber-attacks increases as they can cause significant disruption to society.

9. Globalization: The interconnectedness of the Internet and the ease of cross-border communications make it easy for cybercriminals to launch attacks from anywhere worldwide, making it harder to track and prosecute them.

Despite these limitations, organizations and governments can improve their cybersecurity posture by implementing best practices, investing in advanced technologies, and building a culture of security awareness.

1.2.9 Limitations of outer space security

Outer space security has several limitations that make protecting against threats and hazards challenging. Some of these limitations include:

1. Complexity: The vastness makes monitoring and tracking all objects in orbit, including space debris and satellites, challenging.
2. Technical challenges: The harsh environment of outer space and the limitations of current technology make it challenging to design and operate equipment that can withstand the conditions and provide reliable information.
3. Limited international cooperation: While there are international agreements in place to promote the peaceful use of outer space, there is limited cooperation among nations to address security issues, such as space debris and harmful interference.
4. Limited legal framework: There is a limited legal framework to govern outer space activities, and existing laws and agreements must address security concerns, such as using force and jurisdiction in outer space.
5. Lack of space situational awareness: It is not easy to have accurate and timely information about the location and status of objects in orbit.
6. Limited resources: Smaller organizations, governments, and countries may not have the resources and expertise to defend themselves from space-based threats and hazards.
7. Lack of standardization: There is a lack of standardization in terms of security protocols and lack of international agreements for the responsible use of outer space.
8. Dependence on technology: As the world becomes increasingly dependent on technology, the risk of space-based threats and hazards increases as it can cause significant disruption to society.

Despite these limitations, organizations and governments can improve outer space security by implementing best practices, investing in advanced technologies, and building a culture of security awareness.

1.2.10 Future of cyberspace and outer space

Cybersecurity is constantly evolving as new technologies and threats emerge. Here are a few ways in which cyberspace may evolve in the future:

- Artificial intelligence and machine learning are increasingly used to improve cybersecurity. However, as these technologies advance, attackers may also use them to launch more sophisticated and targeted attacks, making it more challenging to defend against cyber threats.
- Internet of Things (IoT): The number of internet-connected devices is expected to grow, with billions of devices expected to be connected to the Internet in the next few years. This creates new opportunities for attackers to gain access to sensitive data and systems, and it will be increasingly important to secure these devices and to have visibility into the traffic on IoT networks.
- Quantum computing: While this could bring many benefits, it could also make it easier for attackers to break encryption and other security measures currently used to protect data and systems.
- Cloud computing: Cloud computing is becoming more prevalent, with more and more organizations using cloud services to store and process data. This creates new security challenges.
- Cyber warfare: Cyber-attacks are becoming more common and sophisticated, and they are increasingly being used as a tool of warfare. As a result, countries invest more in cyber warfare capabilities and the risks of cyber-attacks being used with other forms of warfare, such as kinetic attacks, increase.

The field of outer space security is constantly evolving as new technologies and threats emerge. Here are a few ways in which outer space security may evolve in the future:

- Space debris: As more and more debris is created in outer space, it will become increasingly important to develop and implement measures to mitigate the risk of collisions and other hazards. This will likely involve increasing space situational awareness capabilities, developing new technologies to remove debris, and promoting international cooperation to address the issue.

- Space mining: As technology advances, it will become increasingly feasible to extract resources from outer space, such as water, minerals, and rare earth elements. This will create new opportunities for economic development and security challenges, as countries and private companies will need to protect their assets and activities in outer space.
- Space weaponization: As countries and private companies invest more in space capabilities, the risk of weaponization of outer space will increase. This will create new security challenges, as countries will need to find ways to prevent outer space weaponization and limit the development, testing, and deployment of weapons in space.
- Cybersecurity in space: As space systems become increasingly dependent on computer systems and networks, the risk of cyber-attacks will increase. Protecting space assets and activities from cyber-attacks will become increasingly important, including jamming signals, spoofing navigation systems, and other forms of cyber-attacks.
- Space governance: As more and more countries and private companies invest in outer space capabilities, there will be an increasing need for international cooperation and governance. This will likely involve the development of new laws and regulations to govern outer space activities and strengthening existing agreements.

1.3 Conclusion

Cyberspace security protects ICT systems and their information from unauthorized access, use, disclosure, disruption, modification, or destruction. Outer space is the physical realm beyond Earth's atmosphere, where satellites, spacecraft, and other objects operate. Outer space security protects space-based assets and the safe and responsible use of outer space. Outer space security concerns include threats such as space debris, space weather, interference of signals and jamming, and space situational awareness, among others. Cyberspace and outer space are related in that they depend on technology and are interconnected. However, they have unique challenges and hazards. Cyberspace security primarily focuses on protecting digital information and communication technology systems and information, while outer space security focuses on the physical security of assets.

The Cybersecurity Kill Chain is primarily used to understand and mitigate cybersecurity threats. While it is primarily used in cybersecurity, it could be adapted and applied to better understand and mitigate the cyber threats to space systems and assets as the integration of space technology with computer systems and networks continues to grow. Cyberspace security has several limitations: complexity, human error, lack of standardization, international agreements, lack of visibility, limited resources, constantly evolving cyber-attacks, dependence on technology, and globalization.

Outer space security has several limitations, such as complexity, technical challenges, limited international cooperation, limited legal framework, lack of space situational awareness, limited resources, lack of standardization, and dependence on technology. Despite these limitations, organizations and governments can improve their cyberspace and outer space security posture by implementing best practices, investing in advanced technologies, and building a culture of security awareness. International cooperation and developing a legal framework that addresses security concerns in cyber and outer space are also required.

Acknowledgments

We thank the Cybersecurity Centre of Excellence – CCoE, Department of Computer Science and Engineering, CMR Institute of Technology, Bengaluru, India. This paper and its research work would not have been possible without it. We would also like to thank the management, principal, vice principal, HoD-CSE, and CMR Institute of Technology, Bengaluru, India, for their constant support and encouragement towards continuous research and development.

References

[1] Smith, J. (2021). Cyber Security Best Practices for Organizations: A Comprehensive Review. Cyber Security Journal.

[2] Robinson, A. (2021). Space Debris: Challenges and Strategies for Mitigation. Journal of Space Environmental Engineering.

[3] Johnson, A. (2020). The Evolution of Cyber Threats: Trends and Mitigation Strategies. Journal of Cyber Security Research.

[4] Patel, K., and Lee, M. (2020). Space Weather and Its Impact on Space Operations. Space Science Review.

[5] Brown, C. (2019). Data Breaches and Cyber Attacks: Lessons Learned and Future Implications. Cyber Defense Journal.

[6] Turner, B. (2019). Satellite Cybersecurity: Protecting Space Assets from Cyber Threats. International Journal of Space Security.

[7] White, M., and Black, L. (2022). Role of Artificial Intelligence in Cyber Security: Opportunities and Challenges. Proceedings of the International Conference on Cyber Security.

[8] Martinez, L., and Davis, R. (2018). Space Situational Awareness: Advancements and Limitations. Proceedings of the International Astronautical Congress.

[9] Green, K. (2018). Cybersecurity Policies and Regulations: A Global Comparative Analysis. Journal of Cyber Law and Policy.

[10] Carter, P. (2020). Outer Space Governance: The Role of International Treaties and Agreements. Journal of Space Law.

[11] Davis, R., and Wilson, S. (2020). Phishing Attacks: Techniques, Detection, and Prevention Measures. Journal of Cybercrime Studies.

[12] King, M., and Brown, C. (2019). Space Traffic Management: Ensuring Safe and Orderly Spaceflight. Space Policy Journal.

[13] Thompson, E. (2019). Ransomware: A Growing Menace in the Cyber Space Landscape. Cyber Threat Analysis Quarterly.

[14] Allen, S., and Turner, D. (2022). Emerging Technologies in Space Security: Opportunities and Risks. Space Technology and Security Conference Proceedings.

[15] Adams, P., and Lee, Q. (2021). Cybersecurity in Critical Infrastructures: Protecting Vital Systems. International Journal of Critical Infrastructure Protection.

[16] Bharathy, A.V., and Karthikeyan, P. (2022). Security and Privacy Policies in Artificially Intelligent 6G Networks: Risks and Challenges. In Challenges and Risks Involved in Deploying 6G and NextGen Networks.

[17] Karthikeyyan, P., Velliangiri, S., and M.I.T. Joseph, S. "Review of Blockchain based IoT application and its security issues," 2019 2nd International Conference on Intelligent Computing, Instrumentation and Control Technologies (ICICICT), Kannur,Kerala, India, 2019, pp. 6–11.

2

Cybersecurity for Space Systems

Galiveeti Poornima and R. Pallavi

Presidency University, India

Email: vlita0502@gmail.com; pallavi.r@presidencyuniversity.in

Abstract

Space security has become increasingly important over the past 10 years as space has become more integrated into our daily lives. However, there is no agreed-upon definition of space security. Recent technological developments like mega-constellations, 5G, the Internet of Things, artificial intelligence, and advanced materials have caused a paradigm shift and transformed the space domain, creating new opportunities and significant challenges; security in space has changed over time. Space security has developed into a sophisticated, multifaceted concept since the Outer Space Treaty (OST) was signed in 1967. The subject dominates the Committee on the Peaceful Uses of Outer Space and its subsidiary bodies' agendas on space law and policy and those of the United Nations General Assembly Committees. In light of the preceding, this chapter seeks to effectively convey the complex idea of space security and summarize the present state.

2.1 Introduction

Space systems encounter a multitude of widely recognized categories of assault, encompassing orbital, kinetic, and electronic warfare, while concurrently exhibiting susceptibility to various manifestations of cyber threat. Cyber-attacks have the potential to manifest in various segments, namely space, communications link, and ground, within the architecture of a space system. Regrettably, these attacks are frequently disregarded in broader deliberations concerning cyber threats targeting critical infrastructure[1].

In relation to the vulnerabilities of space systems and their interconnected links and nodes, it is highly probable that the aforementioned threat will exert a significant impact on the efficacy and/or productivity of concurrent operations conducted by friendly forces. The aforementioned effect shall not solely impact the temporal aspects of planning but may also engender the potentiality for the forfeiture of invaluable military resources and human lives. Space threats encompass a range of potential risks, which may include, but are not confined to, the following:

- The act of tracking and monitoring satellites, along with their corresponding transmissions.
- The utilization of electronic attack (EA) techniques to disrupt or compromise space-based services, targeting the transmission site, the satellite itself, and the user's equipment.
- The occurrence of physical attacks directed toward operational satellites and spacecraft.
- The exploitation of space by hostile entities for the purpose of augmenting their own military capabilities and conducting intelligence activities in preparation for potential conflicts.

The threats above possess the potential to engender communication challenges in the dissemination and implementation of the air tasking order (ATO), impede the practical guidance of weaponry and aircraft towards their intended targets, result in the forfeiture of national overhead and air-breathing intelligence, surveillance, and reconnaissance (ISR) systems, compromise both operations security (OPSEC) and information security (INFOSEC), and influence the state of force protection posture.

The exploitation of vulnerabilities can be achieved through the strategic concentration of attacks on any of the three constituent segments of our space capability, namely the ground, communication (link), and space domains. The ground segment encompasses stationary and mobile terrestrial, maritime or aerial apparatus employed to engage with the space segment. The link segment refers to data transmission between the ground segment, which encompasses terrestrial infrastructure, and the space segment, which contains satellite-based infrastructure. The space segment encompasses various

[1] https://medium.com/the-aerospace-corporation/protecting-space-systems-from-cyber-attack

components, such as satellites, space stations, and reusable space transportation systems. The operational efficacy of our space systems can be enhanced by implementing diverse strategies, encompassing redundancy, hardening, maneuverability, denial, and passive defense.

Given the escalating significance of outer space in contemporary terrestrial existence, there exists a mounting apprehension regarding the potential transformation of the space milieu into a domain rife with hostilities. The modern landscape of military doctrines and strategies progressively acknowledges the significance of space as an operational domain with strategic implications. In certain instances, space is even designated as a domain for engaging in warfare. However, notwithstanding the burgeoning progress in space technology and the escalating dependence on space systems for myriad military and civilian applications, the extant international legal and governance structure governing extraterrestrial endeavors remains scant and seemingly inadequate in its capacity to tackle distinct security apprehensions effectively.

2.2 Background

The assets located in outer space are subject to various space phenomena, including but not limited to orbital debris, as well as the challenging conditions of the external space environment, characterized by extreme temperatures and radiation exposure [1]. Nevertheless, it is imperative to acknowledge that deliberate menaces targeting crucial space systems emanate from nation-states, renegade nation-states, and non-state entities possessing diverse degrees of access to exceedingly proficient and cost-effective anti-satellite (ASAT) armaments [2]. According to Egeli [3], there is a growing concern regarding the vulnerability of critical space infrastructure to ASAT attacks. Additionally, adversaries are actively engaged in developing novel cyber threats, potentially aiming to execute coordinated cyber-storming operations within the realm of "gray-zone" tactics, which fall short of outright warfare [4]. Furthermore, it has been demonstrated that the utilization of kinetic weaponry to obliterate satellites situated in low Earth orbit (LEO) engenders a substantial assemblage of debris, thereby posing a significant hazard to other satellites and engendering a state of contamination within the spatial realm that surpasses the possibility of restoration [3]. According to Bateman's [5] contention, the United States must refrain from using kinetic ASAT measures. On the contrary, the United States must allocate resources toward advancing and implementing non-kinetic armaments, specifically those about electronic or cyber-based methodologies, to neutralize or impair the space assets of their adversaries.

In contradistinction to kinetic armaments, the ramifications of cyber assaults in the extraterrestrial domain remain inadequately comprehended. According to the research conducted by Georgescu et al. [6], it was observed that the air-transport infrastructure exhibited a significant reliance on space systems. These space systems, however, were found to possess a notable likelihood of experiencing spontaneous malfunctions due to their operation within highly demanding environments. In their seminal work, Johnson and Yepez [7] comprehensively analyzed the cyber threats to the global navigation satellite systems (GNSS). The objective of their research is to propose a complete and risk-oriented methodology for the detection and evaluation of potential attack scenarios. This methodology aims to enhance the ability to assess the robustness of safety cases in the face of security threats. The researchers also examined various potential risks posed to GNSS infrastructures, encompassing denial-of-service attacks targeting the ground-based infrastructure, possible compromises to data integrity, and vulnerabilities arising from insider attacks. These threats could undermine GNSS services' accuracy, integrity, availability, and continuity. Nevertheless, their primary emphasis was on safety considerations, neglecting to evaluate cyber risk comprehensively.

The surge in extraterrestrial endeavors primarily stems from the commercial imperative arising from the burgeoning need for space transportation to accommodate novel satellite deployments. The phenomenon under discussion is commonly referred to as "Space 2.0." It encompasses a paradigm shift characterized by the emergence of novel technologies, market participants, launch systems, financing mechanisms, satellite architectures, streamlined small satellite designs, advanced ground antenna systems featuring electronic tracking capabilities, and a notable shift in market dynamics toward networked services [8]. The role of the private sector in outer space activities has assumed a pivotal significance, supplanting the erstwhile dominance of the public sector in driving innovation and technological advancements. According to a recent report by Secure World, it is projected that in the year 2021, a significant number of nations, exceeding 50 in total, will engage in the operation of satellites within the realm of outer space. The emergence of a novel generation of miniature and highly sophisticated satellites engenders the potential peril of spatial congestion and signal encumbrance. The preservation of secure operations about space assets holds paramount importance for the proprietors of said support and the broader domains of global security and peace. This is because space assets play a pivotal role in monitoring anti-terrorist endeavors and surveillance operations and coordinating and managing disaster-related activities [9]. The presence of a substantial reliance on space infrastructure presents a crucial security predicament that critical infrastructure providers, policymakers, and governments often overlook.

Despite its status as a prosperous nation, Australia has made a relatively modest contribution to space policy and the broader domain of space exploration [10]. According to the scholarly work of Georgescu et al. [11], it is imperative to acknowledge the intricate nature of governance in the space domain. This complexity arises from the need for more well-defined jurisdictional boundaries in space, which consequently poses challenges in effectively informing critical national infrastructure processes on our planet. As exemplified by the European Commission, it is posited that space infrastructure assumes a pivotal role as critical infrastructure, upon which indispensable services crucial for the seamless functioning of our societies and economies and the security of our citizens are contingent. The preservation of satellite systems is of utmost importance, and this matter of safeguarding holds significant implications for the European Union that extend well beyond the mere concerns of individual satellite proprietors[2]. Henceforth, it is incumbent upon each nation to discern, designate, and safeguard its respective infrastructure as a critical national infrastructure without a comprehensive regulatory body.

Space technology's pivotal role in commerce and security cannot be overstated; however, it is imperative to acknowledge that it also possesses the potential to be perceived as a menace to the maintenance of international peace [12]. Space exploration has historically been a focal point for governments as the principal patron of extensive research and development endeavors. Consequently, the current scope of international space law does not encompass the space endeavors undertaken by the prominent participants in contemporary space activities, namely the private sector. According to legal provisions, it is mandated that private sector entities operate under the oversight of the state governing their jurisdiction. Furthermore, their operations are expected to align with the pertinent national laws and policies in place. According to Oltrogge et al. [13], numerous states need comprehensive federal legislation on non-space related affairs, resulting in a shortage of more appropriate measures addressing crucial concerns such as liability, technical safeguards, security, and innovation. The imperative for numerous nations to expeditiously establish legislation and harmonize it with the principles of international law is readily apparent. There has been a notable impetus within the global community to cultivate "risk-based" frameworks for the public and

[2] https://op.europa.eu/en/publication-detail/-/publication/ac9cd214-53c6-11ea-aece-01aa75ed71a1/language-en

private sectors [14]. In light of the exponential growth observed in space technology advancements, it is disconcerting to note that the international community of space lawmakers has failed to make substantial progress beyond the initial five treaties and non-binding resolutions established half a century ago.

Regrettably, it is discernible that the United Nations treaties about the realm of extraterrestrial activities need to sufficiently enable the smooth operation of commercial endeavors in the celestial domain [12]. The existing body of space law encompasses a collection of treaties and agreements duly acknowledged and ratified by all participating parties. The lack of universal recognition and acceptance has significantly eroded the universal applicability of space law despite its endorsement through unanimous approval in the United Nations General Assembly [15]. According to Oralova [16], the "Jus Cogens of Space Law" concept pertains to the foundational principles that enjoy unanimous acceptance among the international community as peremptory norms, thereby compelling unwavering adherence from states. The potential deviation from the fundamental principles of non-appropriation of outer space and the freedom to explore and utilize outer space, which is integral to the overarching framework of international space law, is inconsistent with the inherent nature of outer space and has the potential to jeopardize the global space domain and its security. The UNGA Resolution about Legal Principles has duly acknowledged the inherent peril associated with the weaponization of outer space. Nevertheless, the analysis above failed to consider the plausible digital or, as we are presently aware, cyber threats[3]. Furthermore, the resolution in question has yet to undergo any revisions to address the emerging complexities adequately. The United Nations General Assembly's (UNGA) authors focused on the potentialities associated with kinetic assaults, specifically about military endeavors. The North Atlantic Treaty Organization (NATO) has been instrumental in convening deliberations aimed at effectively tackling the multifaceted issues about cyberspace and has subsequently put forth a comprehensive set of suggestions and guidelines, as expounded upon by Tatar [17].

The escalating utilization of space systems for military purposes and many divergent concerns render space assets susceptible to potential attacks. According to the United Nations Office for Outer Space Affairs (UNOOSA), it is stipulated that state parties shall refrain from deploying any entities equipped with nuclear armaments or any other forms of weapons of mass destruction in Earth's orbit, establishing such armaments on celestial bodies

---

[3] https://www.unoosa.org/oosa/en/ourwork/spacelaw/principles/legal-principles.html

or positioning armaments in outer space through alternative means[4]. The Secretary-General of the United Nations Group of Governmental Experts commissioned the study on transparency and confidence-building measures in outer space activities. The research findings have determined that the increasing reliance of the global community on systems and technologies situated in space, along with the consequential information they furnish, necessitates concerted endeavors to mitigate the risks posed to the durability and safeguarding of activities conducted in outer space. The citation provided by the user is in the format of author names followed by the publication year. In an apparent divergence from the extant body of research, governmental entities and industrial sectors have initiated efforts to tackle cybersecurity regarding space infrastructure. In a recent endeavor, the executive branch of the United States government, known as the White House, unveiled a Space Policy Directive (SPD5) titled "Cybersecurity Principals for Space Systems"[5]. This directive establishes a comprehensive framework of guidelines intended to assist space enterprises in formulating their strategies for safeguarding their cyberinfrastructure.

The esteemed organization known as MITRE has recently published a scholarly document titled "Cyber Best Practices for Small Satellites" to effectively respond to the concerns brought forth by SPD5, as articulated by Visner and Kordella [18]. The establishment of a novel Space Information Sharing and Analysis Centre (ISAC) has emerged to foster collaboration within the international space industry. This collaborative effort bolsters our capacity to anticipate, address, and mitigate vulnerabilities, incidents, and threats. Additionally, the ISAC endeavors to facilitate the prompt and practical dissemination of pertinent information among member organizations while assuming the pivotal role of the sector's primary communication conduit about the said report[6]. The ISAC (Information Sharing and Analysis Center) has successfully devised a comprehensive framework to delineate and characterize potential threats within various domains: operations, business systems, providers, and communication segments. In addition to those above, several private entities have disseminated specific scholarly contributions. For instance, the Aerospace Industries Association has released the inaugural National Aerospace Standard NAS9924, designating it as the "Cyber Security Baseline." The Aerospace Industries Association (AIA) has recently formulated a set of fundamental

---

[4] https://www.unoosa.org/oosa/oosadoc/data/documents/2017/stspace/stspace61rev.2 0.html

[5] Directive5, 2020. Cybersecurity Principles for Space Systems. National Space Policy.

[6] https://s-isac.org/

controls with a specific emphasis on bolstering cybersecurity within the supply chain of the aerospace and defense sectors[7].

There has been a growing scholarly fascination with ASAT cyber threats in recent years. This surge in interest can be attributed to the limited comprehension surrounding the strategic, legal, and technical ramifications of such threats, as highlighted by Heinl [19]. Numerous scholars have engaged in a discourse regarding the constraints associated with the "non-binding doctrine" of jus ad Bellum in the context of cyber conflicts, as expounded upon in the notable work known as the "Tallinn Manual" [20, 21, 22]. Hence, the substantial reliance on spatial infrastructure presents a crucial, albeit insufficiently acknowledged, security predicament for critical infrastructure providers and policymakers. The vulnerabilities in necessary space infrastructure manifest in various attacks and armaments commonly called ASAT weapons. As mentioned above, the armaments encompass a range of classifications: kinetic, non-kinetic, electronic, and cyber. In the event of a substantial cyber-attack, the robustness of commercial space infrastructure would probably become fragmented and diminished, necessitating costly replacement measures. Consequently, essential terrestrial services reliant on said infrastructure would experience prolonged unavailability. Hence, a present-day predicament entails the cultivation, imposition, and execution of cybersecurity and resiliency measures for an intricate and substantial space infrastructure encompassing diverse systems, jurisdictions, and legal frameworks. Nevertheless, it is imperative to acknowledge the prevailing shortage of peer-reviewed scholarly investigations about the cybersecurity domain concerning space infrastructure. The extant body of literature remains relatively scant regarding comprehensive examinations of space reference architecture, governance frameworks, and the corresponding tools employed in this context. Consequently, the present study posits that establishing a complete cyber resilience framework is imperative to govern the space industry's burgeoning domain effectively.

2.3 Space Operations

Space operations encompass all activities intended to occur or conducted in the vast expanse of outer space. In pursuit of this objective, terrestrial activities aimed at venturing into outer space, such as a launch, are deemed space operations. From planning, training, and mission control perspectives,

---

[7] https://www.aia-aerospace.org/news/aia-announces-first-cyber-security-standard

this classification remains applicable even in the improbable scenario of an unsuccessful launch. The provided definition elucidates how specific events should be addressed, considering legal and technical perspectives [23].

To ensure the proper functioning of artificial satellites, it is imperative to incorporate three fundamental operational components: space segment, ground segment, and user. The components above are interconnected, forming integral constituents of a broader spatial ecosystem. Collectively, these entities constitute the surface of attack as they engage in collaborative and symbiotic interactions.

2.3.1 Ground segments

The ground segments encompass the entire terrestrial components within the system, including transmitting and receiving stations. The fundamental constituents of the system under consideration surround the ground station, operation center, launch facilities, and integration test facilities. All of the facilities above necessitate the expertise and competence of highly qualified personnel responsible for providing and sustaining operational continuity within the respective domains. Ground segments' de-centralization and global distribution are strategically implemented to serve technical objectives. The continuous establishment of communication between the ground segment infrastructure and the remaining two operational components is imperative to prevent potential misinterpretation of commands and mitigate catastrophic events.

The imperative for the telecommunication infrastructure lies in its inherent need for resilience and perpetual operational functionality. Due to this rationale, many constituents exist on a global scale to mitigate the occurrence of communication blind spots. These may originate from various governmental or non-governmental entities. The European Space Agency (ESA) is effectively leveraging its telecommunication center in the United Kingdom alongside its national centers, including the Italian center in Malindi (Base Broglio) and the French launching site in the French Guiana.

In contemporary times, there has been a notable surge in both the quantity and diversity of cyber events. Examples of such occurrences encompass distributed denial of service (DDoS), ransomware, SQL injection, and brute-force attacks. The ground segment is susceptible to being targeted due to its substantial workforce and the inherent vulnerability of its infrastructures and equipment. This vulnerability may propagate to additional features. Hence, it is imperative to underscore the significance of these segments in fortifying the defensive capabilities, as they serve as a formidable barrier against potential breaches, effectively thwarting any pernicious threats from infiltrating the space segments. Implementing continuous surveillance, proactive measures to alleviate risks, and comprehensive personnel training are imperative to effectively avert and counter potential cyber and physical incidents.

2.3.2 User segments

The user segment encompasses the customer terminals, serving as the primary operational control interface. These terminals facilitate the issuance of commands to the control center and receive the processed data intended for dissemination.

During the nascent stages of the space era, the primary participants consisted of governmental entities propelled by military motivations. Subsequently, the emergence of a market void facilitated the entry of private enterprises, which made substantial investments in this domain.

2.3.3 Space segments

The space segment encompasses the satellite and the associated ground facilities, comprising telemetry and command systems.

The space segments encompass the entire infrastructure that currently resides or is planned for deployment within the spatial domain. Over time, these segments have undergone evolutionary changes in their arrangements and configurations, aiming to fulfill diverse objectives, including serving as launchers, rovers, and modules for the construction of space laboratories and satellites. In the nascent stages of space exploration, a solitary satellite primarily occupied the spatial domain, exemplified by the Envisat. However, as technological progress has unfolded, these satellites have undergone a reduction in physical dimensions while concurrently witnessing a proliferation in quantity. We find ourselves amidst an era characterized by comprehensive constellations, or even colossal constellations, spearheaded by private enterprises such as SpaceX's Starlink initiative.

Satellites exhibit diverse characteristics and serve various purposes, encompassing earth observation, geo-location/navigation, communications, and space exploration[8].

---

[8] https://www.dia.mil/Portals/27/Documents/News/Military Publications/Space Threat V14 020119 sm.pdf

2.4 Cybersecurity

Cybersecurity, or information technology security, encompasses a comprehensive set of safeguards and strategies to safeguard computer systems, networks, software, hardware, and information assets against unauthorized access [24]. Cybersecurity is progressively assuming greater significance due to the amplified dependence on computer systems, smart devices, internet service, wireless networks, intelligent applications, and the multitude of devices that comprise the Internet of Things (IoT). This includes the imperative need for a defensive approach to counteract malicious attacks [25]. Computer systems worldwide require the implementation of detection and predictive analysis systems for cyber threats, which are contingent upon the development of intelligent models. The foundational framework of cybersecurity is predicated upon the principles of integrity, confidentiality, and availability of information, as elucidated in previous scholarly works [26]. Cybersecurity's primary objective is to safeguard cyberspace's integrity, confidentiality, and availability, encompassing a wide range of digital domains, including but not limited to personal information, e-government data, e-marketing websites, e-business reports, and diverse applications, against unauthorized intrusion [27]. Furthermore, this comprehensive study encompasses safeguarding software and equipment, which play a pivotal role in upholding the confidentiality and authenticity of the protected data, thereby shielding it from an array of potential threats and malicious intrusions [28]. Threats encompass the potential perils of computer security that exhibit arduous or elusive characteristics, thereby rendering their detection and mitigation challenging. By their very nature, these threats infringe upon the established security protocols, thereby constituting security violations. Consequently, any endeavor to perpetrate such violations is commonly denoted as an attack [29].

2.5 Categories of Threats to Space Systems

The vulnerabilities posed to space systems can be categorized into four discrete classifications, delineated by the method of assault:

- Kinetic physical
- Non-kinetic physical
- Electromagnetic
- Cyber

The selection of attack methodologies by assailants is contingent upon various factors, including but not limited to the specific target, desired outcomes, and the availability of counter-space technologies. The impacts, reversibility, and traceability of attacks exhibit considerable heterogeneity. The diversity of attack mechanisms is contingent upon the level of awareness of the defender and the general public regarding an impending attack, the attacker's capacity to evaluate the extent of harm inflicted, and the potential for unintended consequences. The subsequent segment elucidates the various manifestations of counter-space attacks that can be executed alongside a compendium of open-source intelligence about the ongoing endeavors of other nations in the domain of development and experimentation.

2.5.1 Kinetic physical

Kinetic physical attacks are endeavors aimed at striking a satellite or triggering the detonation of a warhead near it. On the other hand, non-kinetic physical attacks seek to disrupt or impair the physical functioning of a satellite without engaging in direct physical contact. The 2007 Chinese test of a direct-ascent ASAT weapon is a compelling illustration of the ramifications of kinetic physical assaults. The presence of 35 satellites in LEO, a region predominantly occupied by imaging satellites, renders them susceptible to direct ascent kinetic ASAT weapons, such as those employed in the Chinese test. This vulnerability arises from lower altitude orbits being more accessible to nations with constrained missile capabilities, thereby enabling a more significant number of countries to acquire this capacity. Co-orbital threats pose a significant vulnerability to satellites, as they entail the exploitation of an existing satellite in orbit to launch an attack against another satellite. One illustrative instance pertains to utilizing a space mine, which can secretly track a designated satellite and initiate a controlled explosion upon receiving a specific command. In satellite dynamics, it is worth noting that a co-orbital satellite can engage in a process commonly referred to as satellite capture, whereby it seizes hold of another satellite to affect its displacement or de-arbitration. Furthermore, it can connect with the designated satellite and impede its normal functioning. In addition to its conventional applications, it is worth noting that a satellite can be strategically maneuvered into a crossing orbit, thereby enabling it to intercept a target satellite with minimal advance notice. As previously expounded, the Russian Federation embarked upon the development of co-orbital ASAT armaments commencing in the 1960s. It diligently persisted in conducting rigorous testing of said weaponry throughout the 1970s. China, India, Japan, and various European nations possess the necessary technological capabilities to construct and deploy diminutive satellites for this intended objective, with the potential for additional countries to partake in this endeavor imminently [30]. The contemporary developments observed within space exploration suggest that China is potentially engaged in examining and experimenting with satellites designed for on-orbit servicing and debris removal. Notably, these satellites possess the dual capability of potentially being employed as co-orbital ASAT weapons[9].

Applying kinetic physical attacks often produces profound and harmful consequences for the targeted satellites and ground stations. Furthermore, it is imperative to acknowledge that kinetic assaults executed in the realm of space engender the production of space debris, which possesses the potential to inflict harm upon satellites owned by nations or commercial entities that are not directly embroiled in the conflict above. Kinetic weaponry typically exhibits a discernible attribution, frequently engenders irreversible consequences, and entails a notable propensity for collateral damage. Table 2.1 summarizes threat characteristics vs. kinetic physical.

2.5.2 Non-kinetic physical

Non-kinetic modalities of physical assault possess the potential to exhibit comparable efficacy in terms of satellite disruption, degradation, and destruction while concurrently exhibiting reduced visibility and, in certain instances, heightened challenges in terms of attribution. Directed energy weapons, encompassing laser and high-powered microwave systems, show the capability to expeditiously engage space systems, achieving a prompt response within a matter of seconds. Furthermore, these weapons possess the potential to generate effects that may not be readily discernible to the operator of the satellite, thereby rendering their impact less immediately apparent. Moreover, it is worth noting that directed energy weapons have the potential to be deployed on various platforms such as ships, aircraft, other satellites, or even ground-based installations. Utilizing a high-powered laser, for instance, can engender deleterious effects upon pivotal satellite components, namely solar arrays, through the process of inducing excessive thermal energy within select satellite regions. Moreover, it is worth noting that the utilization of a laser with relatively low power has the potential to induce temporary visual impairment or even permanent ocular damage to the sensors installed on a satellite. The assailant, however, may encounter difficulties in accurately

[9] http://spaceflight101.com/long-march-7-maiden-launch/aolong-1-asat-concerns/.

Table 2.1 Threat characteristics vary by kinetic physical.

		Kinetic physical		
	Types of attacks	**Ground station attack**	**Direct-ascent attack**	**Co-orbital ASAT**
Threat characteristics	**Attribution**	Variable attribution depending on the mode of attack	Launch site can be attributed	Can be attributed by tracking the previously known orbit
	Reversibility	Irreversible	Irreversible	Irreversible or reversible depending on the mode of attack
	Awareness	May or may not be publicly known	Publicly known depending on trajectory	May or may not be publicly known
	Attacker damage assessment	Near real-time confirmation of success	Near real-time confirmation of success	Near real-time confirmation of success
	Collateral damage	Station may control multiple satellites' potential loss of life	Orbital debris could affect other satellites in similar objects	May or may not produce orbital debris

predicting the potential transience or permanence of the repercussions stemming from its offensive actions. Following the execution of an attack, the assailant shall remain uncertain regarding the efficacy of the said attack, owing to the absence of overt external manifestations denoting the malfunctioning of a satellite's sensors.

A set of crucial prerequisites must be met to successfully engage a satellite located in outer space from our terrestrial vantage point. These prerequisites encompass the attainment of a laser beam possessing a superior level of beam quality, the utilization of adaptive optics to compensate for atmospheric disturbances, and the employment of sophisticated pointing control mechanisms to navigate the laser beam through the atmospheric medium effectively. The present technology exhibits a substantial cost burden and lacks extensive accessibility [30]. In September 2006, it was purported that China engaged in the act of illuminating US satellites through the utilization of ground-based lasers. This action potentially aimed to impair or disorientate the satellites, thereby implying that this particular technological capability, although sophisticated, remains within the realm of attainability for prospective adversaries [31]. According to intelligence reports, it has come to light that Russia is currently developing an airborne lazing platform. This particular platform can present challenges in terms of tracking and the ability to effectively target a broader range of orbits within a reasonable timeframe [32].

Nuclear weapons can function as physical armaments through their deployment in space or at elevated altitudes, generating an electromagnetic pulse capable of inflicting detrimental effects upon satellite-based electronics. The above method of assault, nevertheless, exhibits a lack of discrimination in its ramifications as the electromagnetic pulse generated permeates and impacts all satellites within the range of visual perception. Moreover, the residual presence of energetically charged particles resulting from the said occurrence engenders a heightened radiation milieu that influences all satellites within a given orbital domain [33]. The conduct of trials about such armaments has been rendered illegal since the Partial Test Ban Treaty was ratified in 1963. The treaty has garnered signatures and ratifications from over 100 nations, with the notable exceptions of China and North Korea[10]. Table 2.2 summarizes threat characteristics vs. non-kinetic physical.

2.5.3 Electromagnetic

Instead of physically sabotaging space system components, EAs are focused on disrupting the transmission and reception of data by space systems. Jamming refers to an electronic assault wherein radio frequency signals are employed to disrupt and impede the transmission of communications. For a jamming device to effectively disrupt the intended target, it is imperative that the said jammer operates within the same frequency band as the target's antenna. Furthermore, the jammer must be positioned within the field of view of the antenna above. In contrast to physical assaults, jamming possesses the distinct characteristic of being entirely reversible. Upon disengagement of the jamming device, the disrupted communication channels can be effectively reinstated, thereby restoring the flow of information. The utilization of an uplink jammer entails deliberate interference with signals transmitted from a terrestrial source toward a satellite, thereby impeding the satellite's ability to discern between the authentic signal and the extraneous noise generated

[10] http://disarmament.un.org/treaties/t/test ban/text

Table 2.2 Threat characteristics vary by non-kinetic physical.

		Non-kinetic physical			
	Types of attacks	**High alt. nuclear det.**	**High-powered laser**	**Laser dazzling/ blinding**	**High-powered microwave**
Threat characteristics	**Attribution**	Launch site can be attributed	Limited attribution	Clear attribution of the lunar's location at the time of attack	Limited attribution
	Reversibility	Irreversible	Irreversible	Irreversible or reversible, the attacker may not be able to control	Irreversible or reversible, attacker may not be able to control
	Awareness	Publicly known	Only the satellite operator will be aware	Only the satellite operator will be aware	Only the satellite operator will be aware
	Attacker damage assessment	Near real-time confirmation of success	Limited confirmation of success if the satellite begins to drift uncontrolled	Limited confirmation of success	Limited confirmation of success if local RF signals can be monitored
	Collateral damage	Higher radiation levels in the orbit would persist for months or years	Could leave the target satellite disabled and uncontrolled	None	Could leave target satellite disabled and uncontrolled

by the jamming device. The interference of the control link through uplink jamming can effectively impede the transmission of commands from satellite operators to the satellite. However, it is imperative to note that the uplink jammer necessitates a position within the visual range of the antenna on the satellite receiving the command link. Consequently, the jammer must be situated near the command station located on the terrestrial surface [30].

The impact of an uplink jammer on satellite users is characterized by its broad scope, affecting numerous individuals. Conversely, a downlink jammer exhibits a more confined influence, primarily impacting a localized subset of users. The phenomenon of downlink jamming entails the deliberate interference with satellite users by generating extraneous signals within the frequency range of the downlink signal emitted by the said satellite. The efficacy of a downlink jammer is contingent upon its power output, which should be commensurate with the strength of the signal being received on the terrestrial surface. Additionally, the jammer must be situated within the field of view of the receiving terminal's antenna in order to disrupt the signal successfully. The aforementioned constraint serves to restrict the quantity of users who may potentially experience interference from a solitary jamming device. Given that a considerable number of terrestrial terminals employ directional antennas oriented toward the celestial sphere, it is generally imperative for a downlink jamming device to be situated in an elevated position relative to the targeted terminal, in order to effectively disrupt its communication. The aforementioned constraint may be effectively addressed through the utilization of a downlink jammer situated on an aerial or celestial platform, strategically positioned to interpose the jamming device between the terminal and the satellite. Additionally, this functionality enables the jamming device to encompass a broader geographical region, thereby potentially impacting a larger number of individuals utilizing wireless communication services[11]. Ground-based terminals equipped with omnidirectional antennas, such as a multitude of GPS receivers, possess a broader angular coverage, rendering them more vulnerable to downlink jamming originating from various azimuthal directions on the terrestrial plane.

Spoofing is a manifestation of an electronic assault wherein the perpetrator endeavors to replicate a bona fide radio frequency signal to deceive a specific target into erroneously perceiving and engaging with a spurious signal. The potential for an assailant to engage in "downlink spoofing" with regard to satellite communication is a matter of concern. This malicious act

[11] Ibid., 275.

involves the creation of a counterfeit signal, which users inadvertently latch onto, subsequently enabling the injection of fabricated data. In addition to the vulnerabilities above, it is imperative to acknowledge the potential for an assailant to engage in command-and-control uplink signal spoofing, thereby enabling the acquisition of satellite control for malicious intents. The most effective countermeasures against this form of spoofing entail the implementation of signal encryption, as the assailant would be compelled to decipher the encryption to generate a seemingly authentic signal. Additionally, using highly directional antennas that effectively obstruct signals emanating from alternative directions is a noteworthy safeguard.

In 2011, the Islamic Republic of Iran asserted its successful interception of an unmanned aerial vehicle (UAV) operated by the United States of America through electronic warfare techniques, namely jamming and GPS spoofing [34]. Subsequent scholarly investigations have revealed that the integrity of encrypted military global positioning system (GPS) signals can be compromised by utilizing a device that diligently captures the encrypted password and subsequently retransmits it with a marginal temporal discrepancy. This intricate procedure, commonly called "meaconing," has been identified as a potential vulnerability in GPS security[12]. The necessity of decrypting the GPS encryption is obviated in this context, as the rebroadcast signal solely encompasses a temporally delayed replica of the primary signal. Through the gradual manipulation of the temporal delay incorporated, it is plausible to deceive an autopilot system reliant on GPS for navigational purposes into perceiving a state of steady and unvarying horizontal flight, whereas in actuality, the aircraft may be engaged in vertical ascent or descent, or even executing a lateral turn. Table 2.3 summarizes threat characteristics vs. electromagnetic. Table 2.4 summarizes threat characteristics vs. cyber-attacks.

2.5.4 Cyber-attacks

Similar to numerous contemporary military systems, satellites are susceptible to cyber assaults. Cyber-attacks can intercept data, corrupt data, or seize control of methods to fulfill malicious objectives. In similar to numerous contemporary military systems, satellites are susceptible to cyber assaults. Cyber-attacks can intercept data, corrupt data, or seize control of methods to fulfill malicious objectives. In contrast to EAs, which disrupt the transmission

---

[12] http://gpsworld.com/innovation-gnss-spoofing-detection-correlating-carrier-phase-with-rapid-antenna-motion/.

Table 2.3 Threat characteristics vary by electromagnetic.

	Types of Attacks	Electromagnetic Uplink Jamming	Downlink Jamming	Spoofing
Threat characteristics	**Attribution**	Modest attribution depending on the mode of attack	Modest attribution depending on the mode of attack	Modest attribution depending on the mode of attack
	Reversibility	Reversible	Reversible	Reversible
	Awareness	Satellite operator will be aware; public may or may not be	Satellite operator will be aware; public may or may not be	May or may not be known to the public
	Attacker damage assessment	Limited to confirmation of success	Limited confirmation of success if local RF signals can be monitored	Limited confirmation of success if effects are visible
	Collateral damage	Only disrupts the signals targeted and possibly adjacent frequencies	Only disrupts the signals targeted and possibly adjacent frequencies	Only effects the specific RF signals targeted

of radio frequency signals, cyber-attacks specifically focus on compromising the integrity and functionality of data and the underlying systems that rely on said data. Every interface within the system that handles data possesses the potential to be exploited as an intrusion point. This encompasses the antennas on both the satellites and ground stations and the landlines that establish connections between ground stations and terrestrial networks. The ramifications of a cyber-attack on space systems encompass a spectrum of consequences, encompassing data loss, extensive disruptions, and the plausible scenario of irrevocable satellite loss. The results would be twofold in the hypothetical scenario wherein an adversary was to execute a cyber-attack to commandeer a satellite successfully. Firstly, the adversary would be able to disrupt all forms of communication facilitated by the said satellite. Secondly, the adversary could inflict irreparable harm upon the satellite by depleting its propellant reserves or causing detrimental damage to its electronic systems.

Table 2.4 Threat characteristics vary by cyber-attacks.

		Cyber-attacks		
	Types of attacks	**Data intercept/ monitoring**	**Data corruption**	**Seizure of control**
Threat characteristics	**Attribution**	Limited or uncertain attribution	Limited or uncertain attribution	Limited or uncertain attribution
	Reversibility	Reversible	Reversible	Irreversible or reversible depending on the mode of attack
	Awareness	May or may not be known to the public	Satellite operator will be aware; public may or may not be	Satellite operator will be aware; public may or may not be
	Attacker damage assessment	Near real-time confirmation of success	Near real-time confirmation of success	Near real-time confirmation of success
	Collateral damage	None	None	Could have target satellite disabled and uncontrolled

Furthermore, identifying the root cause behind the loss of control of a satellite poses a considerable challenge for controllers, given the intermittent occurrence of inadvertent malfunctions. Attributing a cyber-attack can pose significant difficulties, rendering it a formidable task, if not impossible. This is primarily due to the diverse techniques employed by attackers to obfuscate their true identities, including using compromised servers as launchpads for their malicious activities.

The disclosure made in 2009 brought to light the utilization of readily accessible software by insurgents in Iraq, enabling them to intercept and decipher video transmitted via satellite communication links from surveillance aircraft operated by the United States. This phenomenon was made feasible due to the absence of requisite encryption apparatus in specific US aircraft, allowing the insurgents to access contemporary visual information identical to that being observed by the US military[13]. The U.S.-China Economic

[13] "Insurgents Hack U.S. Drones," The Wall Street Journal, December 17, 2009.

and Security Review Commission has previously provided instances wherein cyber-attacks were employed to target United States government satellites' command and control systems. As per the report published by the Commission in 2011, it is noteworthy that an attack of considerable efficacy was directed toward NASA's Terra EOS satellite in 2008[14]. In June and October, it was reported that unauthorized individuals could assume satellite control for 2 and 9 minutes, respectively. It is worth noting, however, that no commands were executed during these incidents[15].

2.6 Vulnerabilities of Cyber-attacks to Satellites

Satellites, as elucidated by the National Institute of Standards and Technology (NIST), are subject to a level of vulnerability that specific individuals may not fully comprehend. Specifically, NIST defines a cyber-attack as an assault executed through cyberspace, with the explicit objective of impeding, hindering, eradicating, or malevolently manipulating an enterprise's computing environment or infrastructure. Furthermore, such an attack may compromise data integrity or steal classified information.

In light of the general progression of knowledge about space technologies, it is evident that "security by obscurity" is no longer tenable for space systems. This shift can be attributed to satellites' increasing digitization and software-centric nature, which has consequently broadened the potential avenues for attacks. Adversaries may employ many techniques to impede, hinder, eradicate, or malevolently manipulate satellites or their terrestrial systems responsible for satellite command and control. The methodologies encompass a spectrum of cyber-attacks, commencing with unsophisticated "script kid-die" endeavors executed by individuals, denoted as TIER I threats. On the other end of the spectrum, we encounter nation-state-level assaults involving intricate strategies such as supply chain intrusions or space-based attacks. A cyber-attack is not a singular, homogeneous threat. Rather, it manifests in various manifestations, exhibiting many entry points and avenues for exploitation. Once initiated, it can unleash a wide range of debilitating consequences. Figure 2.1 describes the high-level threat vectors and the required level of sophistication to carry out various threats.

---

[14] U.S.-China Economic and Security Review Commission, 2011 Report to Congress, 216.

[15] https://www. geek.com/geek-pick/chinese-hackers-took-control-of-nasa-satellite-for-11-minutes-1442605/.

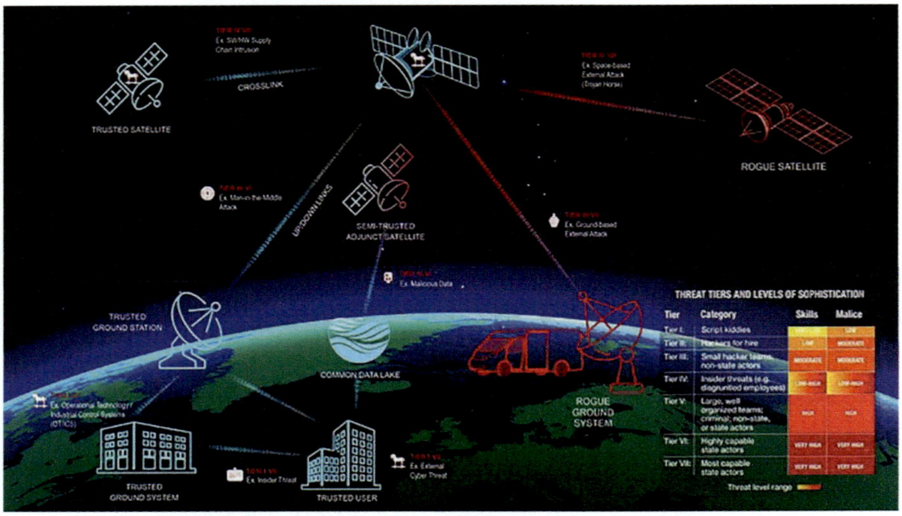

Figure 2.1 Extreme threat vectors and sophistication needed to carry out certain threats[16].

The increasing complexity of space systems has led to their characterization as enigmatic entities, commonly referred to as "black boxes," due to their intricate and interdependent nature within cyberspace. The inherent peril lies in the fact that cyber threats exhibit a comparatively lower development cost when juxtaposed with alternative ASAT technologies. Moreover, it is worth noting that cyber-attacks possess the capability to encompass a substantial attack radius, thereby directing their focus toward an entire assemblage of satellites within a given constellation.

The development and utilization of cyber-attacks hold significant allure for adversaries, especially during periods of conflict. In the context of satellites, it is commonly posited that the distinction lies within the communications link, specifically the radio frequency link or the encompassing ground system. In the event of a breach in the boundary, the satellite lacks sufficient internal safeguards, enabling an adversary to operate without hindrance within the system. This scenario resembles the initial stages of conventional cybersecurity, wherein border firewalls served as the sole means of safeguarding against intrusions. The principles of defense-in-depth are presently

---

[16] https://medium.com/the-aerospace-corporation/protecting-space-systems-from-cyber-attack-3db773aff368

devising contemporary terrestrial information technology (IT) systems of utmost security.

Both conventional expansive developments and contemporary expeditiously evolving space systems must adopt a cyber-hardened design that incorporates defense-in-depth principles comprehensively. Historically, the emphasis on cybersecurity has predominantly revolved around fortifying the terrestrial component. At the same time, scant attention and scholarly inquiry have been directed toward safeguarding the celestial element, namely the satellite. To mitigate the risk of a successful cyber-attack on a satellite, it is imperative to implement cybersecurity measures across a space system's ground and space segments. Doing so can significantly reduce the likelihood of such an attack.

2.7 Space System Attack Strategies

The strategies employed to engage a space system exhibit variability contingent upon the specific entry point or intended target. A space system comprises three fundamental segments: the ground, link, and space segments as shown in Figure 2.2. The strategic maneuvers employed shall vary depending on the specific part subjected to assault.

"Hacking" is employed to delineate a strategic maneuver within the terrestrial domain, wherein conventional cyber methodologies are deemed productive. The ground segment predominantly comprises traditional information technology infrastructure such as Windows and Linux workstations and servers. Consequently, traditional tactics, techniques, and procedures (TTPs) can be effectively employed from the ground system to the modems and antennas. The resemblance between attacks on modems and antennas and those targeting industrial control systems (ICS) is evident. Many of the tactics, techniques, and procedures (TTPs) outlined in MITRE's Adversarial Tactics, Techniques, and Common Knowledge (ATT&CK) framework have demonstrated a notable degree of efficacy when employed against terrestrial systems. The MITRE ATT&CK framework for enterprise and industrial control systems (ICS) can be effectively used to discern vulnerabilities within ground systems.

The employed strategies would be primarily aligned with the adversary's desired outcomes. These strategies may encompass actions such as impeding the reception of ground telemetry and mission data processing, obstructing all forms of communication with the satellite, impeding the ground's capacity to execute satellite commands, diminishing trust in system health assessment tools, compromising mission data integrity, extracting mission data, or even executing commands directly on the satellite.

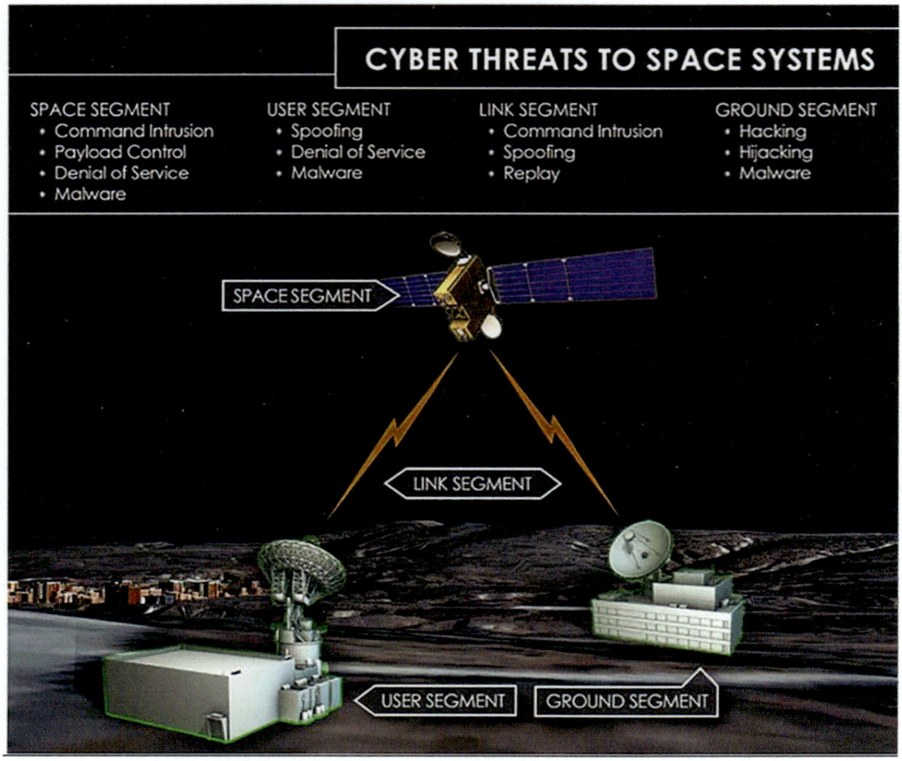

Figure 2.2 Different parts of space systems are attacked differently. Image courtesy of NASIC.

The link segment exhibits a diminished array of tactical options due to its heightened level of comprehension and robust security measures. Satellite providers have traditionally adhered to a well-established convention of safeguarding the link segment through the implementation of communications security (COMSEC), which serves the purpose of upholding the confidentiality of data both when it is stationary and during its transmission. Additionally, transmission security (TRANSEC) measures are employed to guarantee the uninterrupted availability of communications and to restrict intelligence gathering. Without these safeguards, various techniques such as jamming, spoofing, command link instruction, and radio frequency replay attacks can be employed to influence the satellite. Although predominantly characterized by their electronic nature, link layer attacks warrant due consideration in formulating a comprehensive space system strategy.

Regarding the space segment, it is customary for satellites to consist of numerous components that undertake the tasks of data processing, storage, and transmission. It is important to note that each of these components possesses the potential to function as a vulnerable point susceptible to potential attacks. Satellites as shown in Figure 2.3, of embedded hardware and software working within the spatially segregated realm of the cosmos.

During the developmental phase of the system, it is pertinent to acknowledge that it is susceptible to similar vulnerabilities as those encountered by conventional embedded systems. The satellite resembles an Internet of Things (IoT) device within the spatial domain. Undeniably, adversaries are actively searching for ways to exploit residents' vulnerabilities to gain unauthorized access to and manipulate the capabilities of space vehicles. The space vehicle subsystems that are susceptible to cyber-attacks encompass:

- Attitude determination and control (AD&C)
- Command and data handling (C&DH)
- Electrical power and distribution subsystem (EPDS)
- Propulsion subsystem (PS)
- Structures and mechanisms subsystem (SMS)
- Telemetry, tracking, and command (TT&C)
- Thermal control subsystem (TCS)

In satellite targeting, it is reasonable to assert that the flight software assumes a position of heightened significance compared to other subsystems, owing to its integration with the command and data handling (C&DH) system and other subsystems encompassed within the satellite's architectural framework. The pernicious nature of flight software (FSW) based attacks permeates the entirety of the mission's lifecycle. The field switchable waveform (FSW) can be reprogrammed throughout its mission, thereby allowing adversaries ample opportunity to introduce or activate malevolent logic during the developmental phase on terrestrial grounds and subsequent post-launch operations.

Hardware-based attacks are predominantly confined to the pre-launch phase, during which physical access is more readily attainable. Preserving the hardware supply chain or physical integrity during the developmental stage is paramount. However, it is imperative to emphasize that safeguarding the FSW layer, precisely the re-programmable logic embedded within field-programmable gate arrays (FPGAs), necessitates meticulous vigilance throughout the entire lifespan of the space vehicle. Similar to other computational systems, satellites are susceptible to cyber intrusions throughout various stages, including engineering, deployment, and operations. In analyzing

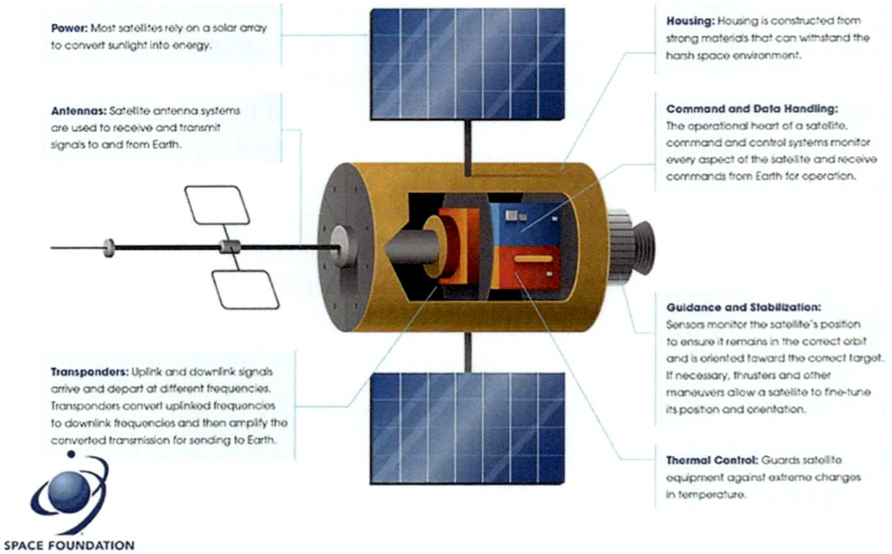

Figure 2.3 Generic satellite parts. Illustration by Space Foundation.

pertinent cyber threats and vulnerabilities, it is imperative to ascertain the objectives an assailant seeks to discern the specific areas and methodologies through which an adversary may direct their efforts toward a satellite.

2.8 Cyber-attacks on Space Assets: Legal Consequences

Numerous discrepancies and policy gaps are evident in implementing cybersecurity measures within space infrastructure, particularly about space-based assets. The delineation of guidelines, standards, and suggested norms is typically constrained within the confines of a specific field or the territorial jurisdiction of a particular state during the negotiation phase.

The successful integration of cybersecurity measures within the space legal framework is contingent upon the field's ability to accommodate and respond to these evolving dynamics effectively. The cyberspace and outer space domains consistently lack distinct, robust, and cohesive incorporation within a legal framework. Space activities should not solely adhere to the principles of space law as a specialized legal framework but should also align with the broader regulations of general international law. The legal framework governing space, cyber-operations, and activities is often fragmented and needs to be more cohesive.

Suppose a state engages in a vicious cyber-operation against another state. In that case, it becomes imperative to ascertain whether it falls short of constituting an employment of force regarding its magnitude and ramifications. By Article 2, Section 4 of the United Nations Charter, it is incumbent upon all member states of the United Nations to exercise restraint in their interactions on the international stage, refraining from employing or even threatening to employ force against any sovereign state's territorial integrity or political independence. Such actions would be deemed incompatible with the fundamental objectives of the United Nations. The interpretation of Article 2, Section 4 of the United Nations Charter suggests that it proscribes the act of making threats or employing force against a sovereign state's territorial integrity or political independence. The binding nature of the United Nations Charter establishes a legal framework that obligates sovereign states. In disputes, it is pertinent to note that only states possess the requisite eligibility to present themselves before the esteemed International Court of Justice (ICJ). Furthermore, it is worth noting that while the ICJ has not yet dealt with any specific cases about cyber incidents, there exist previous cases that possess substantive content that bears relevance to the discourse surrounding cyber law. Without a positive attribution to nation-states, it is imperative to acknowledge that the United Nations Charter lacks applicability, thereby rendering the International Court of Justice devoid of jurisdiction to address matters about individuals or private entities.

The challenge of attributing cyber warfare attacks arises from the inherent difficulty in tracing their origins. Determining the source of a cyber-attack is inherently intricate, primarily owing to the potential presence of an assailant within a distinct jurisdiction.

Within cyberspace, one may duly contemplate the doctrinal elucidation encapsulated within the Tallinn Manual 2.0. A cyber-operation may meet the criteria for being considered a use of force that constitutes an act of aggression, as per the International Court of Justice's application of the concept of "scale and effects" in determining specific actions constituting an armed attack [35]. The concept of "scale and effects" encompasses various components [36], the presence of which can be utilized to classify a cyber-operation as an employment of force [21]. The Lotus principle may disregard legal constraints in cyberspace, contingent upon the physical ramifications of an evil cyber-operation. Moreover, it is worth noting that the OST of 1967 specifically proscribes the deployment of armaments possessing immense destructive capabilities within outer space. The absence of a precise definition for the concept of cyber-attack results in a potential lack of applicability of international obligations to states engaged in malevolent cyber-operations against other states. Hence, it can be inferred that they are not explicitly proscribed. Space assets that utilize cyber technologies are susceptible to various threats and necessitate robust protective measures. The interpretation of the concept of "peaceful use of outer space" varies among states, with some perceiving it as denoting a "non-military" approach. In contrast, others construe it as signifying a lack of aggression. The absence of a consensus on this matter presents a significant challenge, as it creates ambiguity regarding the appropriate course of action for states targeted by malicious cyber-operations. Specifically, it raises questions about how these states can effectively defend themselves and respond in a manner that aligns with the principles of international law.

The International Court of Justice (ICJ) has deliberated upon the notion that actions of a non-kinetic nature may be deemed as constituting the utilization of force [35]. Moreover, it is worth noting that the ICJ deliberated upon the matter in a 1996 advisory opinion, wherein it concluded that universally recognized principles and norms of international humanitarian law apply to all manifestations of armed conflict, including those occurring in the realms of outer space and cyberspace [37]. The inquiry into the status of the various stakeholders is also subject to scrutiny. In the context of non-conflict periods, it is plausible for an evil entity to initiate an assault on computer systems that comprise the space infrastructure by utilizing malicious software. In this particular scenario, the presence of an actor operating either autonomously or under the auspices of a state, an international entity, or a military force remains uncertain. Therefore, ascertaining the actor's military or civilian status presents a considerable challenge, as does identifying the applicable legal framework for the present case.

Discussions surrounding the characteristics and utilization of malevolent cyber endeavors within non-armed conflict contexts, alongside the extent and ramifications of cyber criminality, have engendered scholarly discourse. Pertinent guidance about these matters can be gleaned from soft law. At present, there are ongoing endeavors encompassing autonomous initiatives aimed at formulating comprehensive guidelines that delineate the legal framework governing the realm of cyberspace. These initiatives include the Tallinn Manual 2.0, which seeks to explain the pertinent international laws governing cyberspace, and initiatives focused on military space operations. The ongoing dialogues occurring beyond formal multilateral channels' confines serve as a valuable source of ideas and perspectives. Two prominent initiatives, namely the MILAMOS Project and the Woomera manual, represent noteworthy non-governmental endeavors aimed at formulating comprehensive guidelines elucidating the application of international law to military operations conducted in outer space, encompassing both periods of peace and armed conflict. The capacity to delineate international norms and standards about global conduct constitutes a distinct manifestation of soft power, thereby underscoring the significance of states' engagement in such deliberations. The advent of novel extraterrestrial endeavors has engendered the emergence of developing economic interests. There has been a discernible surge in the propensity of states to exhibit a heightened concern for safeguarding the interests of private enterprises and organizations. Regrettably, our current state of affairs appears to lack a robust and cohesive international legal framework. Lawmakers face a formidable task in comprehending the indispensability of a strong cybersecurity strategy encompassing space operations and digital systems. Cyber-attacks can engender significant disruptions that undermine the operational capabilities of private enterprises, governmental entities, and military forces, impeding their ability to provide robust services. The consideration of the international dimension and the involvement of numerous companies and organizations in the supply chain for the construction of space assets constitute a significant facet to be taken into account. The security issue holds paramount importance about the critical space infrastructure among all the involved parties. The existence of interdependence within the supply chain engenders a situation wherein the weakness of a single actor or state has ramifications for the efficacy of other actors and forms in addressing cyber threats. Due to the inherent difficulty in establishing a robust regulatory framework upheld by a formidable governing body, implementing mitigation measures aimed at averting the worsening of cyber vulnerabilities may represent the most productive approach to safeguarding the interests of space stakeholders. By adopting such measures, these stakeholders can comprehensively assess the potential risks associated with compromising their infrastructure through hostile cyber intrusions. The primary concern faced by actors within the commercial sector pertains to the imperative task of upholding an adequate degree of cybersecurity, particularly in instances where the satellite serves as a critical infrastructure or has implications for national security. Space stakeholders must comprehensively evaluate potential hazards and effectively weigh them against the expenses associated with mitigating said hazards. This assessment process is contingent upon various factors, including the significance of the transmitted data, the value attributed to the information systems, and the critical nature of the methods above.

2.9 Space Systems Cyber-attacks

Various nation-states and criminal organizations have already compromised the integrity of space assets. The attacks that garnered the highest number of references were directed toward space assets supported by governmental and corporate entities. The attacks above are a testament to the insufficiency of cybersecurity measures within adequately financed space initiatives, thereby rendering them vulnerable to malicious intrusions perpetrated by hackers.

2.9.1 IP satellite communication attacks

One of the most intriguing satellite attacks thus far pertained not to the hackers' inclination to compromise the space system but to the technological capabilities facilitated by said space system. Kaspersky Lab has reported the identification of a cyber-espionage entity known as Turla, which is based in Russia. This group has successfully infiltrated a satellite internet service provider to conceal their cyber-espionage endeavors. The targeted countries encompass a broad range, from the United States to the former Eastern Bloc [38]. Through the utilization of a terrestrial antenna, Turla can discern IP addresses emanating from satellite internet users, subsequently initiating a TCP/IP connection originating from the purloined satellite IP address. Turla can obscure its malicious activities by using the stolen intellectual property satellite address. The detectability of the attack is not readily discernible due to the espionage operation's inherent capacity to operate without perceptibly impeding the performance of the unsuspecting user. The efficacy of the attack hinges upon the concurrent utilization of the IP address by both the evil hacker and the legitimate user. Due to the shared IP address between the victim and the attacker's machines, the ensuing attack possesses an inherent stealthiness, rendering it improbable to be detected by intrusion detection systems. Regardless of the degree of stealthiness exhibited by space-based cyber-attacks, their potential to inflict substantial harm upon the operational activities of end-users remains a matter of concern. Consider a hypothetical scenario wherein malicious actors utilize the Turla technique to direct their efforts toward a geographically distant electric substation. The potential for an assailant to intercept either uplink or downlink packets originating from the target's IP address or inject data into the user system linked to the said IP address exists. Introducing fabricated data into a substation component can instigate a state-estimation attack, leading to detrimental consequences such as grid damage and operational downtime.

2.9.2 GPS satellite attacks

A recent incident involving a space-borne cyber assault has resulted in the compromise of GPS systems, which heavily rely on satellite infrastructure to triangulate precise geographical coordinates on our planet. Introducing extraneous signals into the receiver spectrum of the GPS satellite can result in the operational malfunction of a GPS receiver located on Earth, thereby impeding its ability to provide a reading accurately. The methodology above can be identified as the practice of jamming. According to recent reports, it has been observed that the Russian Federation has implemented the deployment of GPS jamming devices on a substantial number of cellular towers, estimated to exceed 250,000 in quantity. The primary objective behind this strategic move is to impede the accurate navigation capabilities of incoming missiles from the United States of America [39]. While it is true that GPS 18 jamming attacks have been previously employed and are not inherently classified as cyber-attacks, it is essential to note that GPS spoofing, on the other hand, falls under the category of cyber-attacks due to its involvement in the deliberate manipulation of the GPS signal. GPS spoofing poses a significantly greater peril than jamming due to its ability to create the illusion of standard GPS functionality. The device's integrity remains intact despite a spoof, rendering its detection arduous and posing a significant threat to critical systems. There exist various methodologies for the act of spoofing a GPS satellite. One potential modus operandi involves compromising the satellite receiver apparatus, thereby manipulating the resultant output signal originating from the satellite. An additional avenue for using the GPS satellite system arises through the implementation of a false data injection attack. In this scenario, an evil entity employs either a GPS signal simulator, which possesses inherent limitations in its ability to deceive the receiver, or a software-defined spoofer consistently. Software-defined spoofers have been found to exhibit a higher degree of reliability in comparison to their conventional counterparts. The operational modality of these systems involves the discreet incorporation of a nearly invisible spurious signal in conjunction with the authentic movement. The power of the counterfeit movement is systematically augmented until it reaches a level at which the recipient perceives the bogus signal as the genuine signal [40]. The construction of a software-defined spoof attack system, exemplified by Professor Todd Humphreys at the respected University of Texas, Austin [41], can be achieved at a relatively modest expense ranging from 1000 to 2000 USD. Attacks employing software-defined spoofing are not hypothetical or executed within controlled laboratory environments. The inaugural instance of a GPS spoofing attack targeting more than 20 vessels in the Black Sea was documented by the U.S. Maritime Administration in the year 2017[17]. The exchange of messages between one of the vessels affected

[17] https://www.marad.dot.gov/msci/alert/2017/ 2017-005a-gps-interference-black-seal

and their central command center indicated that throughout the attack, the GPS position on their navigation tool intermittently indicated a state of "lost GPS fixing position" [42]. During a specific juncture amidst the assault, the falsified geographical coordinates told that the vessel was near the Gelendzhik airport. However, it is imperative to note that the actual position of the ship deviated by a considerable distance of 25 nautical miles from the aforementioned reported location. As per the findings of Resilient Navigation and Timing, a nonprofit organization dedicated to monitoring incidents related to GPS, it has been observed that anecdotal reports of spoofing occurrences are frequently encountered within the maritime boundaries of Russia[18]. There is a prevalent speculation regarding the utilization of a similar modus operandi by the Iranians in their endeavor to apprehend a United States UAV in December 2011 [43]. In September 2011, Iranian researchers purportedly announced their acquisition of a novel methodology for compromising aircraft systems through GPS spoofing. The method above was exemplified by successfully purchasing an American RQ-170 Sentinel UAV by reconfiguring the GPS signal coordinates, thereby redirecting the drone's intended landing site from Afghanistan to Iran [44]. The United States military attributed the acquisition of the drone to a technical malfunction yet failed to provide a comprehensive account of the mechanisms through which the Iranians successfully obtained the drone in an undamaged state [44].

2.9.3 Government satellite and ground space system attacks

Numerous occurrences have been documented wherein space systems under the purview of the United States government have been subjected to cyber intrusions perpetrated by individuals with malicious intent. On October 7, 2007 and July 23, 2008, an Earth observational satellite, under the joint management of the National Aeronautics and Space Administration (NASA) and the United States Geological Survey (USGS), experienced interference for a duration of "12 or more minutes"[19]. In 2011, the U.S.-China Economic and Security Review Commission corroborated that, after the incident, the involved entity failed to execute all requisite measures to attain complete dominion and oversight over the satellite. On June 20, 2008 and October 22, 2008, the Terra EOS AM-1, an Earth observational satellite under the management of the National Aeronautics and Space Administration (NASA),

---

[18] https://nakedsecurity.sophos.com/2017/09/26/ suspected-mass-spoofing-of-ships-gps-in-the-black-seal

[19] https://www.uscc.gov/sites/ default/files/annualreports/annual repoit full 11.pdf

experienced instances of interference lasting 2 and 9 minutes, respectively. In both cases, the malevolent actors successfully attained command and control over the satellites yet refrained from issuing any directives. The aforementioned cyber intrusions have been ascribed to the nation of China, and it is plausible that the primary objectives of these attacks were to undermine the integrity of satellite imagery, impede the transmission of said imagery, or clandestinely extract the images for unauthorized purposes. According to scholarly discourse, it is posited by researchers that the malevolent actors managed to breach a "commercially operated satellite ground station" by leveraging the public internet as a conduit to access the satellite in question. Given the reliance of numerous space systems on commercially operated systems provided by third parties, it is reasonable to surmise that comparable attacks could potentially render other space systems susceptible to similar vulnerabilities. An additional instance of a breach within a government space system occurred in 2011, wherein unauthorized individuals were able to attain complete operational authority over the Jet Propulsion Laboratory (JPL) of the National Aeronautics and Space Administration (NASA) [45]. The accomplishment above was attained through the unauthorized acquisition of more than 150 user credentials belonging to employees of the National Aeronautics and Space Administration (NASA) [45]. Based on the findings presented in the Inspector General's report [45], it has been determined that individuals with complete access to mission-critical systems at the JPL possess the capability to engage in various unauthorized activities. These activities include but are not limited to (1) altering, duplicating, or eradicating sensitive files; (2) manipulating, altering, or erasing user accounts associated with mission-critical JPL systems; (3) uploading malicious software tools to obtain user credentials and compromising other NASA systems illicitly; and (4) tampering with system logs to obfuscate their illicit actions.

The extent of the potential harm incurred during the period of hacker control remains uncertain. However, the implications of this attack's scope are disconcerting, mainly because the JPL is the host institution for the deep space network (DSN) and other vital space-related infrastructures. The instances above represent a limited subset amidst many documented cyber-attacks targeting space assets [46]. In addition to tangible examples of cyber-attacks, scholarly discourse has explored hypothetical scenarios and illustrative examples of attacks targeting space assets, as evidenced by multiple reports [47, 48]. In light of the numerous cybersecurity incidents concerning space assets, one is compelled to inquire about the measures undertaken to fortify the security of said assets.

2.10 Current Methods for Keeping Cybersecurity Safe

Within the space industry community, it is widely recognized that there needs to be more emphasis on cybersecurity. Nevertheless, the reactions and countermeasures employed to address the diverse threats posed by cybersecurity exhibit a considerable degree of variability. The findings of an audit conducted on NASA during the fiscal year 2015 have indicated the need for a comprehensive overhaul of its cybersecurity standards and protocols. The audit above has identified several assaults on NASA's space assets, which were not made known to the general public. These incidents have consequently served as the impetus behind the demand for comprehensive reform[20]. The endeavors undertaken by NASA may not inherently reflect the broader space industry's cybersecurity consciousness and initiatives. Nevertheless, relatively minor entities engaged in satellite operations seek NASA guidance in establishing benchmarks and adopting optimal methodologies. The cybersecurity posture of well-established private space enterprises, such as SpaceX or Blue Origin, remains undisclosed in public discourse. There has been a growing demand for increased public discourse regarding the strategic measures that SpaceX and other relevant entities intend to undertake to tackle the intricate cybersecurity domain in forthcoming endeavors [49].

2.10.1 Access control management

NASA has undertaken a series of measures to enhance its space assets' security. Having taken into consideration the statement above, it is evident that there exist significant prospects for enhancement. INASA has recently initiated implementing more stringent access control policies within its network of providers and engineers. One of the programs implemented by the National Aeronautics and Space Administration (NASA) is the initiative known as "Spot the Phishing Email"[21]. This measure will be a preventive measure against the previously encountered phishing attacks targeting NASA personnel, which aim to illicitly acquire credentials and gain unauthorized access to valuable intellectual assets.

2.10.2 Specialized security workforce

Furthermore, it is noteworthy that NASA has established specialized teams within their space asset development centers to safeguard their mission

[20] https://oig.nasa.gov/NASA2015ManagementChal1enges.pdf

[21] https://www.nasa.gov/pdf/666064main ITTa1k JUL2012 fma1.pdf

systems. Historically, the Office of the Chief Information Officer (OCIO) vested the overarching responsibility for cybersecurity within NASA. Nevertheless, the OCIO teams needed to help allocate their undivided attention to safeguarding the server infrastructure security of NASA's laboratories, the development and operational infrastructure associated with NASA's mission systems, and the mission systems per se. In response to this matter, NASA's JPL established the Cyber Defense Engineering and Research Group (CDER). The primary objective of the CDER is to specifically cater to mission systems, such as the Mars Science Laboratory or the Europa Clipper. These systems typically possess distinct cybersecurity requisites that differ from those of conventional firewalled data servers. The establishment of specialized teams having particular expertise in mission systems facilitates the provision of tailored analysis and protection for space assets, surpassing the capabilities of conventional security teams responsible for safeguarding servers and data. The primary objective of the CDER is to cultivate tools and methodologies that possess broad applicability across various mission systems, thereby effectuating a reduction in both costs and the complexity of security operations.

2.10.3 Employing appropriate security tools

In a recent development, the National Aeronautics and Space Administration (NASA) has initiated the implementation of encryption protocols for safeguarding data during storage and transmission processes. In the concluding months of 2016, AT&T undertook the encryption of NASA's DSN. This pivotal communication infrastructure underpins NASA's deep space missions, encompassing the deployment of rovers and landers on Mars and the exploration of the solar system's outer reaches by spacecraft[22]. By the preceding section's elucidations regarding the subordinate status of space assets in the realm of cybersecurity, it is noteworthy that AT&T, in response to a report divulging techniques for infiltrating the Mars Rover that surfaced on the internet, proceeded to encrypt the DSN. Encryption facilitates secure communication by ensuring that the content of messages remains confidential and accessible solely to individuals possessing the requisite cryptographic key[23]. The utilization of encryption serves as an initial barrier of protection against malicious actors seeking to usurp the DSN or engage in surveillance

[22] https://www.computerworl d.com/artic le/ 31J0973/space-technology/nasa-installs-vpn-to-protect-deep-space-network.html

[23] https://www.extremetech.com/extreme/134334-could-you-hack-into-mars-rover-curiosity

of transmissions traversing this extensive long-range communication infrastructure, which boasts a substantial monetary investment.

The current encryption of DSN communications raises questions regarding the extent to which other public and private satellites have adopted this security measure. In space asset management, it is observed that the community employs security techniques deemed "relevant" by the developers, resulting in a diverse range of encryption practices. Certain satellites currently use the most recent iteration of the National Institute of Standards and Technology's Advanced Encryption Standard (AES), while alternative satellites implement their proprietary cryptographic schemes [50]. One illustrative instance of a satellite diverging from the conventional AES and venturing into uncharted territory regarding space asset security is represented by a Chinese satellite that employs quantum key distribution for encrypted communication [51]. The development above represents a significant leap forward, surpassing the limitations of previous quantum essential distribution methods, which were confined to distances of only a few hundred kilometers. While employing intricate encryption measures for numerous space assets may be deemed excessive, it is evident that the feasibility of implementing sophisticated security techniques for satellites exists.

2.10.4 Fostering a security culture

The CDER is actively establishing a security culture characterized by prudent digital skepticism within its JPL organizational unit. This endeavor is being initiated by implementing a playful activity known as "Donuts." In instances where a member of the CDER Group inadvertently neglects to secure their computer by leaving it unlocked, it has been observed that a fellow CDER teammate seizes the opportunity to exploit this vulnerability by sending an email from the said unlocked computer, with the subject line containing the term "Donuts." If this correspondence is dispatched, it is imperative for the computer user whose system has been compromised to procure a batch of confectionery items, colloquially referred to as "donuts," for the collective benefit of the team. The team diligently maintains a record of the number of donuts owed by each team member, thereby establishing a system of motivation for team members to ensure the security of their computer systems by locking them when they temporarily vacate their workstations. Initially, the team was receiving a regular allocation of donuts daily. However, as the collective consciousness regarding security measures expanded, the customary provision of donuts ceased. The game in question has undergone expansion into additional domains within the JPL, thereby contributing to the establishment of a security culture characterized by a perpetual state of cybersecurity vigilance.

2.10.5 Developing custom security tools

In conjunction with its efforts to cultivate a cybersecurity culture, the CDER is concurrently engaged in developing novel technological solutions to mitigate the escalating cyber risks encountered by the mission systems of NASA JPL. The team has undertaken many projects, one of which involves creating a tool called the Cyber Analysis Visualization Environment (CAVE). This tool serves the purpose of facilitating the modeling and visualization of potential threats that can spread across a mission system's network [52]. This tool can develop and enhance their comprehension of established vulnerability information about their software, which holds utmost significance in accomplishing the mission.

2.10.6 Engaging with the security research community

The CDER Group at NASA JPL is collaborating with the esteemed Massachusetts Institute of Technology (MIT) to undertake security assessments through educational exercises for students, specifically focusing on mission system software [53]. By enhancing the level of engagement with the broader security research community, a significant potential exists for substantial improvement in the security of mission systems about space assets. While there are advantages to engaging external researchers, it is essential to acknowledge the potential challenge that organizations, such as the JPL clan, may face regarding heightened security risks. These risks arise from the security posture of their research collaborators who possess access to data, networks, or systems. Like NASA, the private space asset industry is enhancing its security measures. However, as previously alluded to, evaluating numerous private sector entities remains unfeasible due to their lack of transparency concerning their cyber-security endeavors. Penetration testers, ethical hackers, and security researchers consistently identify vulnerabilities within diverse satellite network systems, prompting responsible entities to rectify said vulnerabilities. Regrettably, the notifications about vulnerabilities frequently encounter disregard because of the manufacturers' limited capacity to redress them or their skepticism toward the hackers. The intricacies of the life cycle and the accompanying liability concerns previously examined further complicate the process of remedying vulnerabilities. In the event of being disregarded, ethical hackers typically adhere to responsible reporting protocols, divulging the identified vulnerability to the public after a designated timeframe following vendor notification. By publicly disclosing the threat, the ethical hackers aim to elicit widespread awareness regarding the problem at hand and compel the vendor to rectify the issue.

In contrast to GPS satellites, which can be detected and subjected to penetration testing even without direct access to the space asset, the security of other private industry space assets remains inaccessible, thereby precluding evaluation by the security community. SpaceX, Virgin Galactic, and other private entities engaged in developing, owning, and operating space assets exhibit a reluctance to provide unfettered access to their technological advancements for scrutiny by security researchers. This apprehension can be attributed to their preoccupation with rival entities' potential compromise of their proprietary code or confidential information. An additional factor pertains to the concerns of private space asset developers regarding the possible discoveries made by security researchers, which, if divulged to the public, possess the capacity to impact their respective enterprises detrimentally. Moreover, it is worth noting that these companies do not have any obligatory obligation to disclose or report on their cybersecurity testing procedures. Consequently, this lack of transparency poses a significant challenge for the security community in assessing the level of cybersecurity exhibited by these space assets. Once more, it is imperative to acknowledge the prominence of the vulnerability disclosure liability and the associated risks entailed in releasing technology for testing. These factors serve as significant impediments to achieving transparency in the realm of space asset security.

Both public and private space asset organizations have encountered significant challenges in their cybersecurity endeavors, as exemplified by the NASA JPL's commendable efforts in fostering a robust security culture. Nevertheless, it is imperative to acknowledge the existence of significant disparities in the security posture of space assets when juxtaposed with other sectors of critical infrastructure, necessitating prompt attention and resolution. The procedural measures for enhancing performance and efficacy differ between private industry entities and government organizations, as expounded upon in the ensuing discourse.

2.11 The Impact of Dual-use Space Assets on Outer Space Security

The coexistence of dual-use capabilities has been evident since the inception of space exploration, owing to the inherent interconnectedness between the advancement of space launch vehicles and ballistic missile technologies. The increasing dependence of states on satellites, which fulfill both civilian and military roles, has been propelled by the rise and expansion of commercial entities operating in space. Furthermore, the task of discerning the true nature of military capabilities exhibited by satellites poses a considerable challenge in ascertaining whether said capabilities are primarily oriented toward offensive or defensive purposes. Although most assets currently present in orbit lack inherent suitability for weaponization, it is essential to acknowledge the existence of significant ambiguity surrounding the definition of dual-use capabilities and the various applications of dual-use technologies. This ambiguity consequently grants space actors substantial latitude in their actions.

The verification of potential behavioral restrictions necessitates the diligent monitoring of space activities and the enhancement of space situational awareness (SSA). States must thoroughly reassess previous propositions, notably the PAXSAT initiative advanced by Canada[24], as well as the proposal put forth by France about establishing an autonomous international monitoring agency[25]. The proposition entails the establishment of an agency that emulates the Preparatory Commission for the Comprehensive Nuclear-Test-Ban Treaty Organization (CTBTO). This agency would be endowed with space assets owned and operated by a diverse array of states. The primary objective of this agency would be to monitor weapons-related activities. By implementing this initiative, it is posited that an autonomous and universally accessible repository of SSA data could be established. Implementing verifiable measures to constrain activities of concern represents a potentially constructive stride toward realizing the PAROS objective. The presence of non-governmental actors in space introduces distinct challenges, primarily stemming from the dual-use capabilities and dual utilization of space assets. In addition to the potential accidental harm and service disruption that non-governmental space assets may encounter due to debris associated with hostile activities in space, it is crucial to acknowledge the possibility

[24] The PAXSAT project (which stands for Peace Satellite) examined the feasibility of using satellite technology to verify a ban on space weapons. See P. Gasparini Alves, Prevention of an Arms Race in Outer Space (PAROS): A Guide to the Discussions in the Conference on Disarmament, UNI-DIR, 1991, https://www.unidir.org/files/publications/pdfs/prevention-of-an-arms-race-in-outer-space-a-guide-to-the-discussions-in-the-cd-en-451.pdf.

[25] During the first Special Session of the General Assembly on Disarmament, France proposed the creation of an international satellite monitoring agency that would participate in monitoring the implementation of international disarmament and security agreements, as well as in investigating specific situations. See General Assembly, "Monitoring of Disarmament Agreements and Strengthening of International Security", Report of the Secretary-General, 27 August 1979, https://undocs.org/A/34/374, pp. 12–13.

of deliberately targeting these assets during armed conflicts in the same domain. Instead, objects attain targetable military objectives based on their inherent characteristics, geographical placement, intended function, or actual utilization[26]. Therefore, it can be inferred that any celestial entity fulfilling both military and civilian roles is deemed an army target and consequently falls within the legal purview of military targeting. Given the circumstances, the various entities within the industry possess a significant stake in maintaining a state of tranquility within the domain of space, thereby actively endorsing endeavors that promote the Prevention of an Arms Race in Outer Space (PAROS). According to the principles outlined in the OST, states must assume responsibility for the actions carried out by non-governmental entities in outer space. The experts participating in the discussion emphasized that the most effective approach to safeguarding the interests of the space industry lies in the engagement of binding treaty negotiations and diplomatic endeavors at the domestic level.

An additional domain of apprehension pertains to military capacities encompassing both offensive and defensive capabilities, exemplified by the presence of anti-ballistic missile (ABM) systems equipped with space-based interceptors that possess the potential to be employed as ASAT weaponry. The positioning of strike capabilities within the orbital realm or the deployment of bodyguard systems, which consist of clusters of smaller satellites capable of strategic maneuvering to safeguard specific space entities, is anticipated to be construed as an aggressive action and regarded as contentious right from the initiation of their deployment. The systems above assume a position of utmost significance as primary targets in the eventuality of a conflict, and it is worth noting that they have already played a role in exacerbating tensions, potentially resulting in the swift deterioration of space-based relations. Therefore, significant attention must be devoted to examining and exploring weapon capabilities before further pursuing development.

2.12 Verification Mechanisms: Technology's Role in Space Security Compliance

Notwithstanding inherent constraints, technological advancements, particularly within SSA, facilitate a more lucid and intricate comprehension of space endeavors, particularly in the geosynchronous orbit. By harnessing the capabilities of these dynamic technologies, it becomes feasible to gain an enhanced

[26] https://ihl-databases.icrc.org/ihl/COM/470750067?OpenDocument

understanding of the behaviors exhibited by individuals within the spatial domain. Consequently, this facilitates the cultivation of trust and enables more efficient validation of specific categories of accords about space security.

The current landscape presents a promising environment for exploring opportunities in the social security administration (SSA) field in the verification context. However, it is imperative to acknowledge that these prospects are accompanied by numerous formidable obstacles that demand careful consideration and strategic navigation. The primary block pertains to the escalating intricacy of outer space endeavors, which encompass a growing multitude of objects and diverse classifications alongside innovative modes of operation. Another challenge is the presence of ambiguity, which manifests not only in the inherent dual-use characteristics of things but also in the diverse interpretations that can be derived from the acquired data. One additional challenge that must be acknowledged is the presence of uncertainty. Recognizing that all sensors possess inherent limitations and can only provide partial data is essential. This is primarily due to the need for comprehensive and accurate models that can effectively monitor every object within the vast expanse of outer space. One of the primary challenges, which holds significant magnitude, pertains to the validation of methodologies and data, as well as the imperative task of securing the active involvement of states in verification mechanisms.

In light of the circumstances above, it is imperative for pertinent stakeholders to not only ascertain the accuracy of data quality assessments but also depend on a multitude of data sources. Furthermore, stakeholders must augment their analysis by incorporating supplementary sources of information. Satellite sensors can furnish data about an entity's physical, functional, and operational attributes and capacities. However, it is plausible for stakeholders to employ alternative sources and methodologies rooted in the social sciences to enhance their ability to ascertain intent and other pertinent behaviors within outer space. Recent advancements in natural language processing and the study of human behavior offer valuable insights into discerning whether disparities in behavior stem from cultural discrepancies in the understanding and interpretation of norms, guidelines, and fundamental concepts within regulatory frameworks, as opposed to deliberate malicious intentions.

The indispensability of cooperation and coordination within verification mechanisms cannot be overstated. Establishing efficient collaboration for SSA infrastructure is imperative, not solely for technological and security rationales but also to mitigate the exorbitant expenses associated with sensors and guarantee efficacy by encompassing a broader geographical scope. The successful execution of this endeavor necessitates the active engagement of a diverse array of stakeholders. The European Space Surveillance and Tracking Programme is a potential exemplar. This program effectively disseminates SSA data using a specialized catalog while furnishing operational SSA services to a noteworthy cohort of 130 European organizations[27].

The potential progression of the international community lies in adopting a distributed framework for data collection and compilation, as opposed to relying on a solitary repository or source. Such an approach could engender a heightened sense of trust in the data that has been aggregated. Participatory networks constructed in a grassroots manner exemplify a burden-sharing paradigm wherein the distinct capacities of diverse actors are mutually reinforcing. States or non-state actors lacking SSA capabilities can still participate and acquire these capabilities by making specific contributions in other domains, such as algorithm development, applied mathematics, or decision-making.

A pressing and timely prospect within the realm of SSA pertains to the capacities exhibited by commercial entities. One illustrative instance involves the United States-based enterprise LeoLabs, which is augmenting a worldwide assemblage of synchronized sensors. This concerted effort seeks to facilitate the expeditious comprehension of entities and activities transpiring within LEO. The utilization of commercial capabilities has the potential to address the limitations in data availability, thereby facilitating the identification of historical behavioral patterns and providing valuable insights for governmental deliberations on space trends. Consequently, this knowledge can significantly influence the formulation and implementation of space governance measures.

In light of a prospective forthcoming treaty, it is imperative to underscore the necessity of preemptively instituting transparency mechanisms before embarking upon any endeavor to verify discrete actions about a given event. There exists a requirement for a governing structure that possesses the capacity to quantitatively assess, comprehend, and effectively administer the intricate and densely populated SSA framework. In the interim, states must manifest their commitment by providing substantiation of their endeavors to enhance the transparency, predictability, and accountability of space activities among all relevant entities.

2.13 Conclusion

Cybersecurity for space systems is of paramount importance in our increasingly interconnected and technology-driven world. Space systems play a

[27] See EU SST https://www.eusst.eu/.

critical role in various aspects of modern life, including communication, navigation, weather forecasting, national security, and scientific research. Protecting these systems from cyber threats is essential for ensuring the continued functionality and security of space-based infrastructure.

References

[1] Muhammad Mateen Yaqoob, Muhammad Nazir, Abdullah Yousafzai, Muhammad Amir Khan, Asad Ali Shaikh, Abeer D Algarni, and Hela Elmannai. Modified artificial bee colony based feature optimized federated learning for heart disease diagnosis in healthcare. Applied Sciences, 12(23):12080, 2022.

[2] Wanyong Qiu, Kun Qian, Zhihua Wang, Yi Chang, Zhihao Bao, Bin Hu, Björn W Schuller, and Yoshiharu Yamamoto. A federated learning paradigm for heart sound classification. In 2022 44th Annual International Conference of the IEEE Engineering in Medicine & Biology Society (EMBC), pages 1045–1048. IEEE, 2022.

[3] Lerina Aversano, Mario Luca Bernardi, Marta Cimitile, Martina Iammarino, Debora Montano, and Chiara Verdone. Using machine learning for early prediction of heart disease. In 2022 IEEE International Conference on Evolving and Adaptive Intelligent Systems (EAIS), pages 1–8. IEEE, 2022.

[4] Roohallah Alizadehsani, Moloud Abdar, Mohamad Roshanzamir, Abbas Khosravi, Parham M Kebria, Fahime Khozeimeh, Saeid Nahavandi, Nizal Sarrafzadegan, and U Rajendra Acharya. Machine learning- based coronary artery disease diagnosis: A comprehensive review. Computers in biology and medicine, 111:103346, 2019.

[5] Subhi J Al'Aref, Khalil Anchouche, Gurpreet Singh, Piotr J Slomka, Kranthi K Kolli, Amit Kumar, Mohit Pandey, Gabriel Maliakal, Alexander R Van Rosendael, Ashley N Beecy, et al. Clinical applications of machine learning in cardiovascular disease and its relevance to cardiac imaging. European heart journal, 40(24):1975–1986, 2019.

[6] Sadman Sakib, Mostafa M Fouda, Zubair Md Fadlullah, Khalid Abualsaud, Elias Yaacoub, and Mohsen Guizani. Asynchronous federated learning-based ecg analysis for arrhythmia detection. In 2021 IEEE International Mediterranean Conference on Communications and Networking (MeditCom), pages 277-282. IEEE, 2021. 282. IEEE, 2021.

[7] Shahid Munir Shah and Rizwan Ahmed Khan. Secondary use of electronic health record: Opportunities and challenges. IEEE access, 8:136947–136965, 2020.

[8] Joo Hun Yoo, Ha Min Son, Hyejun Jeong, Eun-Hye Jang, Ah Young Kim, Han Young Yu, Hong Jin Jeon, and Tai-Myoung Chung. Personalized federated learning with clustering: Non-iid heart rate variability data application. In 2021 International Conference on Information and Communication Technology Convergence (ICTC), pages 1046–1051. IEEE, 2021.

[9] Brendan McMahan, Eider Moore, Daniel Ramage, Seth Hampson, and Blaise Aguera y Arcas. Communication-efficient learning of deep networks from decentralized data. In Artificial intelligence and statistics, pages 1273–1282. PMLR, 2017.

[10] Qinbin Li, Zeyi Wen, Zhaomin Wu, Sixu Hu, Naibo Wang, Yuan Li, Xu Liu, and Bingsheng He. A survey on federated learning systems: Vision, hype and reality for data privacy and protection. IEEE Transactions on Knowledge and Data Engineering, 2021.

[11] Stacey Truex, Nathalie Baracaldo, Ali Anwar, Thomas Steinke, Heiko Ludwig, Rui Zhang, and Yi Zhou. A hybrid approach to privacy-preserving federated learning. In Proceedings of the 12th ACM workshop on artificial intelligence and security, pages 1–11, 2019.

[12] Nicola Rieke, Jonny Hancox, Wenqi Li, Fausto Milletari, Holger R Roth, Shadi Albarqouni, Spyridon Bakas, Mathieu N Galtier, Bennett A Landman, Klaus Maier-Hein, et al. The future of digital health with federated learning. NPJ digital medicine, 3(1):119, 2020.

[13] Theodora S Brisimi, Ruidi Chen, Theofanie Mela, Alex Olshevsky, Ioannis Ch Paschalidis, and Wei Shi. Federated learning of predictive models from federated electronic health records. International journal of medical informatics, 112:59–67, 2018.

[14] Akis Linardos, Kaisar Kushibar, Sean Walsh, Polyxeni Gkontra, and Karim Lekadir. Federated learning for multi-center imaging diagnostics: a simulation study in cardiovascular disease. Scientific Reports, 12(1):3551, 2022.

[15] Liang Zou, Zexin Huang, Xinhui Yu, Jiannan Zheng, Aiping Liu, and Meng Lei. Automatic detection of congestive heart failure based on multiscale residual unet++: From centralized learning to federated learning. IEEE Transactions on Instrumentation and Measurement, 72:1–13, 2022.

[16] Rushabh Shah and Alina Chircu. Iot and ai in healthcare: A systematic literature review. Issues in Information Systems, 19(3), 2018.

[17] Alfonso Panarello, Nachiket Tapas, Giovanni Merlino, Francesco Longo, and Antonio Puliafito. Blockchain and iot integration: A systematic survey. Sensors, 18(8):2575, 2018.

[18] Manisha Bahl, Regina Barzilay, Adam B Yedidia, Nicholas J Locascio, Lili Yu, and Constance D Lehman. High-risk breast lesions: a machine learning model to predict pathologic upgrade and reduce unnecessary surgical excision. Radiology, 286(3):810–818, 2018.

[19] Federico Cugurullo. Exposing smart cities and eco-cities: Frankenstein urbanism and the sustainability challenges of the experimental city. Environment and Planning A: Economy and Space, 50(1):73–92, 2018.

[20] Federico Cugurullo. Frankenstein urbanism: Eco, smart and autonomous cities, artificial intelligence and the end of the city. Routledge, 2021.

[21] Ivo Sérgio Guimarães Brites, Lídia Martins da Silva, Jorge Luis Victória Barbosa, Sandro José Rigo, Sérgio Duarte Correia, and Valderi Reis Quietinho Leithardt. Machine learning and iot applied to cardiovascular diseases identification through heart sounds: A literature review. In Informatics, volume 8(4), page 73. MDPI, 2021.

[22] Cameron R Olsen, Robert J Mentz, Kevin J Anstrom, David Page, and Priyesh A Patel. Clinical applications of machine learning in the diagnosis, classification, and prediction of heart failure. American Heart Journal, 229:1–17, 2020.

[23] Federico M Asch, Nicolas Poilvert, Theodore Abraham, Madeline Jankowski, Jayne Cleve, Michael Adams, Nathanael Romano, Ha Hong, Victor Mor-Avi, Randolph P Martin, et al. Automated echocardiographic quantification of left ventricular ejection fraction without volume measurements using a machine learning algorithm mimicking a human expert. Circulation: Cardiovascular Imaging, 12(9):e009303, 2019.

[24] Joon-myoung Kwon, Kyung-Hee Kim, Ki-Hyun Jeon, Sang Eun Lee, Hae-Young Lee, Hyun-Jai Cho, Jin Oh Choi, Eun-Seok Jeon, Min-Seok Kim, Jae-Joong Kim, et al. Artificial intelligence algorithm for predicting mortality of patients with acute heart failure. PloS one, 14(7):e0219302, 2019.

[25] Jeffrey J Nirschl, Andrew Janowczyk, Eliot G Peyster, Renee Frank, Kenneth B Margulies, Michael D Feldman, and Anant Madabhushi. A deep-learning classifier identifies patients with clinical heart failure using whole-slide images of h&e tissue. PloS one, 13(4):e0192726, 2018.

[26] Hadi Habibzadeh and Tolga Soyata. Toward uniform smart healthcare ecosystems: A survey on prospects, security, and privacy considerations. Connected health in smart cities, pages 75–112, 2020.

[27] Saba Latif and Nazir Ahmad Zafar. A survey of security and privacy issues in iot for smart cities. In 2017 Fifth International Conference on Aerospace Science & Engineering (ICASE), pages 1–5. IEEE, 2017.

[28] Guoxin Huang, Yajun Fang, Xiushi Wang, Yu Pei, and Berthold Horn. A survey on the status of smart healthcare from the universal village perspective. In 2018 4th International Conference on Universal Village (UV), pages 1–6. IEEE, 2018.

[29] Hamidreza Arasteh, Vahid Hosseinnezhad, Vincenzo Loia, Aurelio Tommasetti, Orlando Troisi, Miadreza Shafie-khah, and Pierluigi Siano. Iot-based smart cities: A survey. In 2016 IEEE 16th international conference on environment and electrical engineering (EEEIC), pages 1–6. IEEE, 2016.

[30] Saeed H Alsamhi, Ou Ma, Mohammad Samar Ansari, and Faris A Almalki. Survey on collaborative smart drones and internet of things for improving smartness of smart cities. Ieee Access, 7:128125–128152, 2019.

[31] Wei Li, Yuanbo Chai, Fazlullah Khan, Syed Rooh Ullah Jan, Sahil Verma, Varun G Menon, fnm Kavita, and Xingwang Li. A comprehensive survey on machine learning-based big data analytics for iot-enabled smart healthcare system. Mobile networks and applications, 26:234–252, 2021.

[32] Jeevan Kharel, Haftu T Reda, and Soo Y Shin. An architecture for smart health monitoring system based. Journal of Communications, 12(4):228–233, 2017.

[33] Khan Muhammad, Salman Khan, Javier Del Ser, and Victor Hugo C De Albuquerque. Deep learning for multigrade brain tumor classification in smart healthcare systems: A prospective survey. IEEE Transactions on Neural Networks and Learning Systems, 32(2):507–522, 2020.

[34] Safeeullah Soomro, Mahdi H Miraz, Anupama Prasanth, and Mirza Abdullah. Artificial intelligence enabled iot: Traffic congestion reduction in smart cities. NA, 2018.

[35] Zaheer Allam and Zaynah A Dhunny. On big data, artificial intelligence and smart cities. Cities, 89:80–91, 2019.

[36] Umer Majeed, Latif U Khan, Ibrar Yaqoob, SM Ahsan Kazmi, Khaled Salah, and Choong Seon Hong. Blockchain for iot-based smart cities: Recent advances, requirements, and future challenges. Journal of Network and Computer Applications, 181:103007, 2021.

[37] Federico Cugurullo. Urban artificial intelligence: From automation to autonomy in the smart city. Frontiers in Sustainable Cities, 2:38, 2020.

[38] Zaib Ullah, Fadi Al-Turjman, Leonardo Mostarda, and Roberto Gagliardi. Applications of artificial intelligence and machine learning in smart cities. Computer Communications, 154:313–323, 2020.

[39] Sudan Jha, Lewis Nkenyereye, Gyanendra Prasad Joshi, and Eunmok Yang. Mitigating and monitoring smart city using internet of things. Computers, Materials & Continua, 65(2):1059–1079, 2020.

[40] Sushil Kumar Singh, Jeonghun Cha, Tae Woo Kim, and Jong Hyuk Park. Machine learning based distributed big data analysis framework for next generation web in iot. Computer Science and Information Systems, 18(2):597–618, 2021.

[41] Sushil Kumar Singh, AE Azzaoui, Tae Woo Kim, Yi Pan, and Jong Hyuk Park. Deepblockscheme: A deep learning-based blockchain driven scheme for secure smart city. Human-centric Computing and Information Sciences, 11(12):1–13, 2021.

[42] Wuxiong Zhang, Weidong Fang, Qianqian Zhao, Xiaohong Ji, and Guoqing Jia. Energy efficiency in internet of things: An overview. Computers, Materials & Continua, 63(2), 2020.

[43] Nina Cvar, Jure Trilar, Andrej Kos, Mojca Volk, and Emilija Stojmenova Duh. The use of iot technology in smart cities and smart villages: Similarities, differences, and future prospects. Sensors, 20(14):3897, 2020.

[44] Andrea Zanella, Nicola Bui, Angelo Castellani, Lorenzo Vangelista, and Michele Zorzi. Internet of things for smart cities. IEEE Internet of Things journal, 1(1):22–32, 2014.

[45] Leonardo Guevara and Fernando Auat Cheein. The role of 5g technologies: Challenges in smart cities and intelligent transportation systems. Sustainability, 12(16):6469, 2020.

[46] Dr S Smys, Dr Haoxiang Wang, and Dr Abul Basar. 5g network simulation in smart cities using neural network algorithm. Journal of Artificial Intelligence and capsule networks, 3(1):43–52, 2021.

[47] Fabian Dembski, Uwe Wossner, Mike Letzgus, Michael Ruddat, and Claudia Yamu. Urban digital twins for smart cities and citizens: The case study of herrenberg, germany. Sustainability, 12(6):2307, 2020.

[48] David N Ford and Charles M Wolf. Smart cities with digital twin systems for disaster management. Journal of management in engineering, 36(4):04020027, 2020.

[49] Arunmozhi Manimuthu, Venugopal Dharshini, Ioannis Zografopoulos, MK Priyan, and Charalambos Konstantinou. Contactless technologies for smart cities: big data, iot, and cloud infrastructures. SN computer science, 2(4):334, 2021.

[50] Christian Nasulea and Stelian-Mihai Mic. Using blockchain as a platform for smart cities. Journal of E-Technology, 9(2):37–43, 2018.

[51] PS Aithal et al. Blockchain technology: a driving force in smart cities development. International Journal of Applied Engineering and Management Letters (IJAEML), 4(2):237–252, 2020.

[52] Carmen Rotuna, Alexandru GHEORGHITĂ, Alin Zamfiroiu, and Dragoș SMADA ANAGRAMA. Smart city ecosystem using blockchain technology. Informatica Economica, 23(4), 2019.

[53] Juan Manuel Corchado Rodríguez et al. Efficiency and reliability in bringing ai into transport and smart city solutions. 2020.

[54] Jiandong Cao. Research on urban intelligent traffic monitoring system based on video image processing. International Journal of Signal Processing, Image Processing and Pattern Recognition, 9(6):393–406, 2016.

[55] Zuzana Sitavancova and Martin Hajek. Thematic research summary: Intelligent transport systems. 2010.

[56] John F Gilmore and Khalid J Elibiary. Ai in advanced traffic management systems. Association for the Advancement of Artificial Intelligence (AAAI) Technical Report WS-93-04, 1993.

[57] Bilal Ghazal, Khaled ElKhatib, Khaled Chahine, and Mohamad Kherfan. Smart traffic light control system. In 2016 third international conference on electrical, electronics, computer engineering and their applications (EECEA), pages 140–145. IEEE, 2016.

[58] Taqi Tahmid and Eklas Hossain. Density based smart traffic control system using canny edge detection algorithm for congregating traffic information. In 2017 3rd International Conference on Electrical Information and Communication Technology (EICT), pages 1–5. IEEE, 2017.

[59] Saqib Hakak, Wazir Zada Khan, Gulshan Amin Gilkar, Muhammad Imran, and Nadra Guizani. Securing smart cities through blockchain technology: Architecture, requirements, and challenges. IEEE Network, 34(1):8–14, 2020.

[60] Junfeng Xie, Helen Tang, Tao Huang, F Richard Yu, Renchao Xie, Jiang Liu, and Yunjie Liu. A survey of blockchain technology applied to smart cities: Research issues and challenges. IEEE communications surveys & tutorials, 21(3):2794–2830, 2019.

[61] Kichan Nam, Christopher S Dutt, Prakash Chathoth, and M Sajid Khan. Blockchain technology for smart city and smart tourism: latest trends and challenges. Asia Pacific Journal of Tourism Research, 26(4):454–468, 2021.

[62] Simona Ibba, Andrea Pinna, Matteo Seu, and Filippo Eros Pani. Citysense: blockchain-oriented smart cities. In Proceedings of the XP2017 Scientific Workshops, pages 1–5, 2017.

[63] Jeong Hoon Jo, Pradip Kumar Sharma, Jose Costa Sapalo Sicato, and Jong Hyuk Park. Emerging technologies for sustainable smart city network security: Issues, challenges, and countermeasures. Journal of Information Processing Systems, 15(4):765–784, 2019.

[64] Bhuvana Reddy, PS Aithal, et al. Blockchain as a disruptive technology in healthcare and financial services-a review based analysis on current implementations. 2020.

[65] Silvia Liberata Ullo and Ganesh Ram Sinha. Advances in smart environment monitoring systems using iot and sensors. Sensors, 20(11):3113, 2020.

[66] Bartosz Przydatek, Dawn Song, and Adrian Perrig. Sia: Secure information aggregation in sensor networks. In Proceedings of the 1st international conference on Embedded networked sensor systems, pages 255–265, 2003.

[67] Hedde HWJ Bosman, Giovanni Iacca, Arturo Tejada, Heinrich J Wörtche, and Antonio Liotta. Ensembles of incremental learners to detect anomalies in ad hoc sensor networks. ad hoc networks, 35:14–36, 2015.

[68] Lavish Kansal, Gurjot Singh Gaba, Naveen Chilamkurti, and Byung-Gyu Kim. Efficient and robust image communication techniques for 5g applications in smart cities. Energies, 14(13):3986, 2021.

[69] Pavel Hamet and Johanne Tremblay. Artificial intelligence in medicine. Metabolism, 69:S36–S40, 2017.

[70] Andreas Holzinger, Georg Langs, Helmut Denk, Kurt Zatloukal, and Heimo Müller. Causability and explainability of artificial intelligence in medicine. Wiley Interdisciplinary Reviews: Data Mining and Knowledge Discovery, 9(4):e1312, 2019.

[71] Sam Musa. Smart cities-a road map for development. IEEE Potentials, 37(2):19–23, 2018.

[72] Auriol Degbelo, Carlos Granell, Sergio Trilles, Devanjan Bhattacharya, Sven Casteleyn, and Christian Kray. Opening up smart cities: citizen-centric challenges and opportunities from giscience. ISPRS International Journal of Geo-Information, 5(2):16, 2016.

[73] Tuba Bakıcı, Esteve Almirall, and Jonathan Wareham. A smart city initiative: the case of barcelona. Journal of the knowledge economy, 4:135–148, 2013.

[74] Stéphane Roche, Nashid Nabian, Kristian Kloeckl, and Carlo Ratti. Are 'smart cities' smart enough. In Global geospatial conference, pages 215–235, 2012.
[75] Zhihan Lv, Tengfei Yin, Xiaolei Zhang, Houbing Song, and Ge Chen. Virtual reality smart city based on webvrgis. IEEE Internet of Things Journal, 3(6):1015–1024, 2016.
[76] David Monaghan, James O'Sullivan, Noel E O'Connor, Bridget Kelly, Olivier Kazmierczak, and Lorraine Comer. Low-cost creation of a 3d interactive museum exhibition. In Proceedings of the 19th ACM international conference on Multimedia, pages 823–824, 2011.
[77] P Caudell Thomas and WM David. Augmented reality: An application of heads-up display technology to manual manufacturing processes. In Hawaii international conference on system sciences, volume 2. ACM SIGCHI Bulletin, 1992.
[78] Paul Milgram and Fumio Kishino. A taxonomy of mixed reality visual displays. IEICE Transactions on Information and Systems, 77(12):1321–1329, 1994.
[79] Leila Alem, Franco Tecchia, and Weidong Huang. Handsonvideo: Towards a gesture based mobile ar system for remote collaboration. In Recent trends of mobile collaborative augmented reality systems, pages 135–148. Springer, 2011.
[80] Marco de Sá, Judd Antin, David Shamma, and Elizabeth F Churchill. Mobile augmented reality: video prototyping. In CHI'11 Extended Abstracts on Human Factors in Computing Systems, pages 1897–1902. 2011.
[81] Apple Newsroom. The future is here: iphone x, 2017.
[82] N Statt. Samsung partners with google to bring arcore platform to galaxy devices. The Verge, 18, 2017.
[83] David W Roberts, John W Strohbehn, John F Hatch, William Murray, and Hans Kettenberger. A frameless stereotaxic integration of computerized tomographic imaging and the operating microscope. Journal of neurosurgery, 65(4):545–549, 1986.
[84] Jiann-Der Lee, Hao-Che Lee, Chung-Hung Hsieh, Chieh-Tsai Wu, and Shin-Tseng Lee. A projection- based medical augmented reality system. In Proceedings of the 8th International Symposium on Visual Information Communication and Interaction, pages 164–165, 2015.
[85] Carling L Cheung, Chris Wedlake, John Moore, Stephen E Pautler, and Terry M Peters. Fused video and ultrasound images for minimally invasive partial nephrectomy: a phantom study. In Medical Image

Computing and Computer-Assisted Intervention–MICCAI 2010: 13th International Conference, Beijing, China, September 20-24, 2010, Proceedings, Part III 13, pages 408–415. Springer, 2010.

[86] Shun'ichi Tano, Keisuke Suzuki, Kenji Miki, Natsuko Watanabe, Mitsuru Iwata, Tomohiro Hashiyama, Junko Ichino, and Ken Nakayama. Simple augmented reality system for 3d ultrasonic image by see- through hmd and single camera and marker combination. In Proceedings of 2012 IEEE-EMBS Interna- tional Conference on Biomedical and Health Informatics, pages 464–467. IEEE, 2012.

[87] artificial intelligence-and-machine-learning-software-medical-device, 2019. https://www.fda.gov/ medicalΘdevices/software-medical-device-samd/ [Accessed: 2019 April].

[88] Antonio Clim, R˘azvan Zota, Radu Constantinescu, and Iulian Ilie-Nemedi. Health services in smart cities: Choosing the big data mining based decision support. International Journal of Healthcare Management, 13(1):79–87, 2020.

[89] Sheetal Menon and Karuna Jain. Blockchain technology for transparency in agri-food supply chain: Use cases, limitations, and future directions. IEEE Transactions on Engineering Management, 2021.

[90] Raja Wasim Ahmad, Khaled Salah, Raja Jayaraman, Ibrar Yaqoob, Mohammed Omar, and Samer Ellahham. Blockchain-based forward supply chain and waste management for covid-19 medical equipment and supplies. Ieee Access, 9:44905–44927, 2021.

[91] N Nasurudeen Ahamed, TK Thivakaran, and P Karthikeyan. Perishable food products contains safe in cold supply chain management using blockchain technology. In 2021 7th International Conference on Advanced Computing and Communication Systems (ICACCS), volume 1, pages 167–172. IEEE, 2021.

[92] Shaoyong Guo, Yao Dai, Song Guo, Xuesong Qiu, and Feng Qi. Blockchain meets edge computing: Stackelberg game and double auction based task offloading for mobile blockchain. IEEE Transactions on Vehicular Technology, 69(5):5549–5561, 2020.

[93] Xiaohong Zhang and Xiaofeng Chen. Data security sharing and storage based on a consortium blockchain in a vehicular ad-hoc network. Ieee Access, 7:58241–58254, 2019.

[94] Yan Zhuang, Lincoln R Sheets, Yin-Wu Chen, Zon-Yin Shae, Jeffrey JP Tsai, and Chi-Ren Shyu. A patient-centric health information exchange framework using blockchain technology. IEEE journal of biomedical and health informatics, 24(8):2169–2176, 2020.

[95] J Indumathi, Achyut Shankar, Muhammad Rukunuddin Ghalib, J Gitanjali, Qiaozhi Hua, Zheng Wen, and Xin Qi. Block chain based internet of medical things for uninterrupted, ubiquitous, user-friendly, unflappable, unblemished, unlimited health care services (bc iomt u 6 hcs). IEEE Access, 8:216856– 216872, 2020.

[96] Sanjit Kumar Dash, Subasish Mohapatra, and Prasant Kumar Pattnaik. A survey on applications of wireless sensor network using cloud computing. International Journal of Computer Science & Emerging Technologies, 1(4):50–55, 2010.

[97] Jonathan Kissi, Baozhen Dai, Courage SK Dogbe, Jonathan Banahene, and Oyeh Ernest. Predictive factors of physicians' satisfaction with telemedicine services acceptance. Health informatics journal, 26(3):1866–1880, 2020.

[98] Kazem Sohraby, Daniel Minoli, and Taieb Znati. Basic wireless sensor technology. 2007.

[99] Francesca Gherardi and Laura Aquiloni. Sexual selection in crayfish: A review. New Frontiers in Crustacean Biology, pages 213–223, 2011.

[100] Khalid Haseeb, Zahoor Jan, Fahad A Alzahrani, and Gwanggil Jeon. A secure mobile wireless sensor networks based protocol for smart data gathering with cloud. Computers & Electrical Engineering, 97:107584, 2022.

[101] Bernardita Calzon. 24 examples of big data analytics in healthcare that can save people, 2023. https://www.datapine.com/blog/big-data-examples-in-healthcare/ [Accessed: Jun 16th 2023].

[102] Michael Grieves and John Vickers. Digital twin: Mitigating unpredictable, undesirable emergent behavior in complex systems. Transdisciplinary perspectives on complex systems: New findings and approaches, pages 85–113, 2017.

[103] Abdulmotaleb El Saddik. Digital twins: The convergence of multimedia technologies. IEEE multimedia, 25(2):87–92, 2018.

[104] Hawazin Faiz Badawi and Abdulmotaleb El Saddik. Biofeedback in healthcare: State of the art and meta review. Connected Health in Smart Cities, pages 113–142, 2020.

[105] Miltiadis D Lytras and Anna Visvizi. Who uses smart city services and what to make of it: Toward interdisciplinary smart cities research. Sustainability, 10(6):1998, 2018.

[106] Miltiadis D Lytras, Paraskevi Papadopoulou, and Akila Sarirete. Smart healthcare: emerging technolo- gies, best practices, and sustainable policies. Innovation in health informatics, pages 3–38, 2020.

[107] Fedwa Laamarti, Hawazin Faiz Badawi, Yezhe Ding, Faisal Arafsha, Basim Hafidh, and Abdulmotaleb El Saddik. An iso/ieee 11073 standardized digital twin framework for health and well-being in smart cities. IEEE Access, 8:105950–105961, 2020.

[108] Jamal Maktoubian and Keyvan Ansari. An iot architecture for preventive maintenance of medical devices in healthcare organizations. Health and Technology, 9:233–243, 2019.

[109] Pantea Keikhosrokiani. Perspectives in the development of mobile medical information systems: Life cycle, management, methodological approach and application. Academic Press, 2019.

[110] Pantea Keikhosrokiani, Norlia Mustaffa, and Nasriah Zakaria. Success factors in developing iheart as a patient-centric healthcare system: A multi-group analysis. Telematics and Informatics, 35(4):753–775, 2018.

[111] Pantea Keikhosrokiani, Norlia Mustaffa, Nasriah Zakaria, and Rosni Abdullah. Assessment of a medical information system: the mediating role of use and user satisfaction on the success of human interaction with the mobile healthcare system (iheart). Cognition, Technology & Work, 22:281–305, 2020.

[112] Pantea Keikhosrokiani, Norlia Mustaffa, Nasriah Zakaria, and Muhammad Imran Sarwar. Wireless positioning techniques and location-based services: a literature review. Multimedia and Ubiquitous Engineering: MUE 2013, pages 785–797, 2013.

[113] Virginia Smith, Chao-Kai Chiang, Maziar Sanjabi, and Ameet S Talwalkar. Federated multi-task learning. Advances in neural information processing systems, 30, 2017.

[114] Yue Zhao, Meng Li, Liangzhen Lai, Naveen Suda, Damon Civin, and Vikas Chandra. Federated learning with non-iid data. arXiv preprint arXiv:1806.00582, 2018.

[115] Keith Bonawitz, Vladimir Ivanov, Ben Kreuter, Antonio Marcedone, H Brendan McMahan, Sarvar Patel, Daniel Ramage, Aaron Segal, and Karn Seth. Practical secure aggregation for privacy-preserving machine learning. In proceedings of the 2017 ACM SIGSAC Conference on Computer and Communications Security, pages 1175–1191, 2017.

[116] Robin C Geyer, Tassilo Klein, and Moin Nabi. Differentially private federated learning: A client level perspective. arXiv preprint arXiv:1712.07557, 2017.

[117] Mi Luo, Fei Chen, Pengxiang Cheng, Zhenhua Dong, Xiuqiang He, Jiashi Feng, and Zhenguo Li. Metas- elector: Meta-learning

for recommendation with user-level adaptive model selection. In Proceedings of The Web Conference 2020, pages 2507–2513, 2020.

[118] Li Li, Yuxi Fan, Mike Tse, and Kuo-Yi Lin. A review of applications in federated learning. Computers & Industrial Engineering, 149:106854, 2020.

[119] Subrato Bharati, M Mondal, Prajoy Podder, and VB Prasath. Federated learning: Applications, chal- lenges and future directions. International Journal of Hybrid Intelligent Systems, 18(1-2):19–35, 2022.

[120] Payman Mohassel and Yupeng Zhang. Secureml: A system for scalable privacy-preserving machine learning. In 2017 IEEE symposium on security and privacy (SP), pages 19–38. IEEE, 2017.

[121] Niki Kilbertus, Adrià Gascon, Matt Kusner, Michael Veale, Krishna Gummadi, and Adrian Weller. Blind justice: Fairness with encrypted sensitive attributes. In International Conference on Machine Learning, pages 2630–2639. PMLR, 2018.

[122] Dan Bogdanov, Sven Laur, and Jan Willemson. Sharemind: A framework for fast privacy-preserving computations. In Computer Security-ESORICS 2008: 13th European Symposium on Research in Com- puter Security, Málaga, Spain, October 6–8, 2008. Proceedings 13, pages 192–206. Springer, 2008.

[123] Payman Mohassel and Peter Rindal. Aby3: A mixed protocol framework for machine learning. In Proceedings of the 2018 ACM SIGSAC conference on computer and communications security, pages 35–52, 2018.

[124] Toshinori Araki, Jun Furukawa, Yehuda Lindell, Ariel Nof, and Kazuma Ohara. High-throughput semi-honest secure three-party computation with an honest majority. In Proceedings of the 2016 ACM SIGSAC Conference on Computer and Communications Security, pages 805–817, 2016.

[125] Payman Mohassel, Mike Rosulek, and Ye Zhang. Fast and secure three-party computation: The garbled circuit approach. In Proceedings of the 22nd ACM SIGSAC Conference on Computer and Communications Security, pages 591–602, 2015.

[126] Irene Giacomelli, Somesh Jha, Marc Joye, C David Page, and Kyonghwan Yoon. Privacy-preserving ridge regression with only linearly-homomorphic encryption. In Applied Cryptography and Network Security: 16th International Conference, ACNS 2018, Leuven, Belgium, July 2-4, 2018, Proceedings 16, pages 243–261. Springer, 2018.

[127] Rob Hall, Stephen E Fienberg, and Yuval Nardi. Secure multiple linear regression based on homomorphic encryption. Journal of Official Statistics, 27(4):669–691, 2011.

[128] Valeria Nikolaenko, Udi Weinsberg, Stratis Ioannidis, Marc Joye, Dan Boneh, and Nina Taft. Privacy- preserving ridge regression on hundreds of millions of records. In 2013 IEEE symposium on security and privacy, pages 334–348. IEEE, 2013.
[129] Jiawei Yuan and Shucheng Yu. Privacy preserving back-propagation neural network learning made practical with cloud computing. IEEE Transactions on Parallel and Distributed Systems, 25(1):212–221, 2013.
[130] Qingchen Zhang, Laurence T Yang, and Zhikui Chen. Privacy preserving deep computation model on cloud for big data feature learning. IEEE Transactions on Computers, 65(5):1351–1362, 2015.
[131] Abbas Acar, Hidayet Aksu, A Selcuk Uluagac, and Mauro Conti. A survey on homomorphic encryption schemes: Theory and implementation. ACM Computing Surveys (Csur), 51(4):1–35, 2018.
[132] Yoshinori Aono, Takuya Hayashi, Le Trieu Phong, and Lihua Wang. Scalable and secure logistic regres- sion via homomorphic encryption. In Proceedings of the Sixth ACM Conference on Data and Application Security and Privacy, pages 142–144, 2016.
[133] Miran Kim, Yongsoo Song, Shuang Wang, Yuhou Xia, Xiaoqian Jiang, et al. Secure logistic regression based on homomorphic encryption: Design and evaluation. JMIR medical informatics, 6(2):e8805, 2018.
[134] Cynthia Dwork. Differential privacy: A survey of results. In International conference on theory and applications of models of computation, pages 1–19. Springer, 2008.
[135] Yoshinori Aono, Takuya Hayashi, Lihua Wang, Shiho Moriai, et al. Privacy-preserving deep learn- ing via additively homomorphic encryption. IEEE transactions on information forensics and security, 13(5):1333–1345, 2017.
[136] Eugene Bagdasaryan, Andreas Veit, Yiqing Hua, Deborah Estrin, and Vitaly Shmatikov. How to back- door federated learning. In International conference on artificial intelligence and statistics, pages 2938– 2948. PMLR, 2020.
[137] Luca Melis, Congzheng Song, Emiliano De Cristofaro, and Vitaly Shmatikov. Exploiting unintended feature leakage in collaborative learning. In 2019 IEEE symposium on security and privacy (SP), pages 691–706. IEEE, 2019.
[138] Lili Su and Jiaming Xu. Securing distributed gradient descent in high dimensional statistical learning. Proceedings of the ACM on Measurement and Analysis of Computing Systems, 3(1):1–41, 2019.
[139] Hyesung Kim, Jihong Park, Mehdi Bennis, and Seong-Lyun Kim. Blockchained on-device federated learning. IEEE Communications Letters, 24(6):1279–1283, 2019.

[140] Qiang Yang, Yang Liu, Tianjian Chen, and Yongxin Tong. Federated machine learning: Concept and applications. ACM Transactions on Intelligent Systems and Technology (TIST), 10(2):1–19, 2019.

[141] Subrato Bharati, Prajoy Podder, Dang Ngoc Hoang Thanh, and VB Surya Prasath. Dementia classification using mr imaging and clinical data with voting based machine learning models. Multimedia Tools and Applications, 81(18):25971–25992, 2022.

[142] Subrato Bharati, Prajoy Podder, and M Rubaiyat Hossain Mondal. Diagnosis of polycystic ovary syndrome using machine learning algorithms. In 2020 IEEE region 10 symposium (TENSYMP), pages 1486–1489. IEEE, 2020.

[143] Chuan Ma, Jun Li, Ming Ding, Howard H Yang, Feng Shu, Tony QS Quek, and H Vincent Poor. On safeguarding privacy and security in the framework of federated learning. IEEE network, 34(4):242–248, 2020.

[144] Subrato Bharati and Prajoy Podder. Adaptive papr reduction scheme for ofdm using slm with the fusion of proposed clipping and filtering technique in order to diminish papr and signal distortion. Wireless Personal Communications, 113(4):2271–2288, 2020.

[145] Solmaz Niknam, Harpreet S Dhillon, and Jeffrey H Reed. Federated learning for wireless communications: Motivation, opportunities, and challenges. IEEE Communications Magazine, 58(6):46–51, 2020.

[146] Mohammad Mohammadi Amiri and Deniz Gündüz. Federated learning over wireless fading channels. IEEE Transactions on Wireless Communications, 19(5):3546–3557, 2020.

[147] Ha-Vu Tran, Georges Kaddoum, Hany Elgala, Chadi Abou-Rjeily, and Hemani Kaushal. Lightwave power transfer for federated learning-based wireless networks. IEEE Communications Letters, 24(7):1472–1476, 2020.

[148] Fan Ang, Li Chen, Nan Zhao, Yunfei Chen, Weidong Wang, and F Richard Yu. Robust federated learning with noisy communication. IEEE Transactions on Communications, 68(6):3452–3464, 2020.

[149] Thien Duc Nguyen, Samuel Marchal, Markus Miettinen, Hossein Fereidooni, N Asokan, and Ahmad- Reza Sadeghi. Dïot: A federated self-learning anomaly detection system for iot. In 2019 IEEE 39th International conference on distributed computing systems (ICDCS), pages 756–767. IEEE, 2019.

[150] Stefano Savazzi, Monica Nicoli, and Vittorio Rampa. Federated learning with cooperating devices: A consensus approach for massive iot networks. IEEE Internet of Things Journal, 7(5):4641–4654, 2020.

[151] Mehdi Salehi Heydar Abad, Emre Ozfatura, Deniz Gunduz, and Ozgur Ercetin. Hierarchical federated learning across heterogeneous cellular networks. In ICASSP 2020-2020 IEEE International Conference on Acoustics, Speech and Signal Processing (ICASSP), pages 8866–8870. IEEE, 2020.

[152] Weishan Zhang, Tao Zhou, Qinghua Lu, Xiao Wang, Chunsheng Zhu, Haoyun Sun, Zhipeng Wang, Sin Kit Lo, and Fei-Yue Wang. Dynamic-fusion-based federated learning for covid-19 detection. IEEE Internet of Things Journal, 8(21):15884–15891, 2021.

[153] Rajesh Kumar, Abdullah Aman Khan, Jay Kumar, Noorbakhsh Amiri Golilarz, Simin Zhang, Yang Ting, Chengyu Zheng, Wenyong Wang, et al. Blockchain-federated-learning and deep learning models for covid-19 detection using ct imaging. IEEE Sensors Journal, 21(14):16301–16314, 2021.

[154] M Rubaiyat Hossain Mondal, Subrato Bharati, and Prajoy Podder. Co-irv2: Optimized inceptionresnetv2 for covid-19 detection from chest ct images. PloS one, 16(10):e0259179, 2021.

[155] M Rubaiyat Hossain Mondal, Subrato Bharati, and Prajoy Podder. Co-irv2: Optimized inceptionresnetv2 for covid-19 detection from chest ct images. PloS one, 16(10):e0259179, 2021.

[156] Subrato Bharati, Prajoy Podder, M Mondal, and VB Prasath. Co-resnet: Optimized resnet model for covid-19 diagnosis from x-ray images. International Journal of Hybrid Intelligent Systems, 17(1-2):71–85, 2021.

[157] M Rubaiyat Hossain Mondal, Subrato Bharati, Prajoy Podder, and Priya Podder. Data analytics for novel coronavirus disease. informatics in medicine unlocked, 20:100374, 2020.

[158] M Rubaiyat Hossain Mondal, Subrato Bharati, and Prajoy Podder. Diagnosis of covid-19 using machine learning and deep learning: a review. Current Medical Imaging, 17(12):1403–1418, 2021.

[159] Prajoy Podder, Aditya Khamparia, M Rubaiyat Hossain Mondal, Mohammad Atikur Rahman, and Subrato Bharati. Forecasting the spread of covid-19 and icu requirements. 2021.

[160] Subrato Bharati, Prajoy Podder, and M Rubaiyat Hossain Mondal. Hybrid deep learning for detecting lung diseases from x-ray images. Informatics in Medicine Unlocked, 20:100391, 2020.

[161] Subrato Bharati, Prajoy Podder, M Mondal, and VB Prasath. Medical imaging with deep learning for covid-19 diagnosis: a comprehensive review. arXiv preprint arXiv:2107.09602, 2021.

[162] Subrato Bharati, Prajoy Podder, M Rubaiyat Hossain Mondal, and Niketa Gandhi. Optimized nasnet for diagnosis of covid-19 from lung

ct images. In International Conference on Intelligent Systems Design and Applications, pages 647–656. Springer, 2020.

[163] Subrato Bharati, Prajoy Podder, M Rubaiyat Hossain Mondal, Priya Podder, and Utku Kose. A review on epidemiology, genomic characteristics, spread, and treatments of covid-19. Data Science for COVID- 19, pages 487–505, 2022.

[164] Aditya Khamparia, Rubaiyat Hossain Mondal, Prajoy Podder, Bharat Bhushan, Victor Hugo C de Al- buquerque, and Sachin Kumar. Computational Intelligence for Managing Pandemics, volume 5. Walter de Gruyter GmbH & Co KG, 2021.

[165] Subrato Bharati, MR Hossain Mondal, A Khamparia, R Hossain Mondal, P Podder, B Bhushan, V Al- buquerque, and S Kumar. 12 Applications and challenges of AI-driven IoHT for combating pandemics: a review. De Gruyter Berlin, Boston, 2021.

[166] Karthik V Sarma, Stephanie Harmon, Thomas Sanford, Holger R Roth, Ziyue Xu, Jesse Tetreault, Daguang Xu, Mona G Flores, Alex G Raman, Rushikesh Kulkarni, et al. Federated learning improves site performance in multicenter deep learning without data sharing. Journal of the American Medical Informatics Association, 28(6):1259–1264, 2021.

[167] Subrato Bharati and Prajoy Podder. Performance of cnn for predicting cancerous lung nodules using lightgbm. Artificial Intelligence for Data-Driven Medical Diagnosis, 1:1–18, 1.

[168] Prajoy Podder, Subrato Bharati, and M Rubaiyat Hossain Mondal. 10 automated gastric cancer detection and classification using machine learning. Artificial Intelligence for Data-Driven Medical Diagnosis, De Gruyter, pages 207–224, 2021.

[169] Subrato Bharati, Prajoy Podder, and M Mondal. Artificial neural network based breast cancer screening: a comprehensive review. arXiv preprint arXiv:2006.01767, 2020.

[170] Subrato Bharati, Mohammad Atikur Rahman, and Prajoy Podder. Breast cancer prediction apply- ing different classification algorithm with comparative analysis using weka. In 2018 4th International Conference on Electrical Engineering and Information & Communication Technology (iCEEiCT), pages 581–584. IEEE, 2018.

[171] Subrato Bharati, Prajoy Podder, Rajib Mondal, Atiq Mahmood, and Md Raihan-Al-Masud. Comparative performance analysis of different classification algorithm for the purpose of prediction of lung cancer. In Intelligent Systems Design and Applications: 18th International Conference on Intelligent Systems Design and Applications (ISDA

2018) held in Vellore, India, December 6-8, 2018, Volume 2, pages 447–457. Springer, 2020.

[172] Aditya Khamparia, Subrato Bharati, Prajoy Podder, Deepak Gupta, Ashish Khanna, Thai Kim Phung, and Dang NH Thanh. Diagnosis of breast cancer based on modern mammography using hybrid transfer learning. Multidimensional systems and signal processing, 32:747–765, 2021.

[173] Subrato Bharati. How artificial intelligence impacts businesses in the period of pandemics. Journal of the International Academy for Case Studies, 26(5):1–2, 2020.

[174] Subrato Bharati, Prajoy Podder, and Pinto Kumar Paul. Lung cancer recognition and prediction accord- ing to random forest ensemble and rusboost algorithm using lidc data. International Journal of Hybrid Intelligent Systems, 15(2):91–100, 2019.

[175] Prajoy Podder, Subrato Bharati, Mohammad Atikur Rahman, and Utku Kose. Transfer learning for classification of brain tumor. In Deep learning for biomedical applications, pages 315–328. CRC Press, 2021.

[176] Mohammed Aledhari, Rehma Razzak, Reza M Parizi, and Fahad Saeed. Federated learning: A survey on enabling technologies, protocols, and applications. IEEE Access, 8:140699–140725, 2020.

[177] Subrato Bharati, Prajoy Podder, and Md Raihan-Al-Masud. Eeg eye state prediction and classification in order to investigate human cognitive state. In 2018 International Conference on Advancement in Electrical and Electronic Engineering (ICAEEE), pages 1–4. IEEE, 2018.

[178] Md Sadiq Iqbal, Md Nasim Akhtar, AHM Shahariar Parvez, Subrato Bharati, and Prajoy Podder. En- semble learning-based eeg feature vector analysis for brain computer interface. In Evolutionary Computing and Mobile Sustainable Networks: Proceedings of ICECMSN 2020, pages 957–969. Springer, 2021.

[179] Benjamin S Glicksberg, Kipp W Johnson, and Joel T Dudley. The next generation of precision medicine: observational studies, electronic health records, biobanks and continuous monitoring. Human molecular genetics, 27(R1):R56–R62, 2018.

[180] Peter B Jensen, Lars J Jensen, and Søren Brunak. Mining electronic health records: towards better research applications and clinical care. Nature Reviews Genetics, 13(6):395–405, 2012.

[181] Ziad Obermeyer, Brian Powers, Christine Vogeli, and Sendhil Mullainathan. Dissecting racial bias in an algorithm used to manage the health of populations. Science, 366(6464):447–453, 2019.

[182] Ziad Obermeyer, Brian Powers, Christine Vogeli, and Sendhil Mullainathan. Dissecting racial bias in an algorithm used to manage the health of populations. Science, 366(6464):447–453, 2019.

[183] Dong Yang, Ziyue Xu, Wenqi Li, Andriy Myronenko, Holger R Roth, Stephanie Harmon, Sheng Xu, Baris Turkbey, Evrim Turkbey, Xiaosong Wang, et al. Federated semi-supervised learning for covid region segmentation in chest ct using multi-national data from china, italy, japan. Medical image analysis, 70:101992, 2021.

[184] Rui Duan, Mary Regina Boland, Zixuan Liu, Yue Liu, Howard H Chang, Hua Xu, Haitao Chu, Christo- pher H Schmid, Christopher B Forrest, John H Holmes, et al. Learning from electronic health records across multiple sites: A communication-efficient and privacy-preserving distributed algorithm. Journal of the American Medical Informatics Association, 27(3):376–385, 2020.

[185] Arju Manara Begum, M Rubaiyat Hossain Mondal, Prajoy Podder, and Subrato Bharati. Detecting spinal abnormalities using multilayer perceptron algorithm. In International Conference on Innovations in Bio-Inspired Computing and Applications, pages 654–664. Springer, 2021.

[186] Li Huang, Andrew L Shea, Huining Qian, Aditya Masurkar, Hao Deng, and Dianbo Liu. Patient cluster- ing improves efficiency of federated machine learning to predict mortality and hospital stay time using distributed electronic medical records. Journal of biomedical informatics, 99:103291, 2019.

[187] Ziyi Li, Kirk Roberts, Xiaoqian Jiang, and Qi Long. Distributed learning from multiple ehr databases: contextual embedding models for medical events. Journal of biomedical informatics, 92:103138, 2019.

[188] P Vignesh Raja and E Sivasankar. Modern framework for distributed healthcare data analytics based on hadoop. In Information and Communication Technology: Second IFIP TC5/8 International Conference, ICT-EurAsia 2014, Bali, Indonesia, April 14-17, 2014. Proceedings 2, pages 348–355. Springer, 2014.

[189] Yejin Kim, Jimeng Sun, Hwanjo Yu, and Xiaoqian Jiang. Federated tensor factorization for computational phenotyping. In Proceedings of the 23rd ACM SIGKDD International Conference on Knowledge Discovery and Data Mining, pages 887–895, 2017.

[190] Junghye Lee, Jimeng Sun, Fei Wang, Shuang Wang, Chi-Hyuck Jun, Xiaoqian Jiang, et al. Privacy-preserving patient similarity learning in a federated environment: development and analysis. JMIR medical informatics, 6(2):e7744, 2018.

[191] Dianbo Liu, Dmitriy Dligach, and Timothy Miller. Two-stage federated phenotyping and patient repre- sentation learning. In Proceedings of the conference. Association for Computational Linguistics. Meeting, volume 2019, page 283. NIH Public Access, 2019.

[192] Otkrist Gupta and Ramesh Raskar. Distributed learning of deep neural network over multiple agents. Journal of Network and Computer Applications, 116:1–8, 2018.

[193] Praneeth Vepakomma, Otkrist Gupta, Tristan Swedish, and Ramesh Raskar. Split learning for health: Distributed deep learning without sharing raw patient data. arXiv preprint arXiv:1812.00564, 2018.

[194] Sabri Boughorbel, Fethi Jarray, Neethu Venugopal, Shabir Moosa, Haithum Elhadi, and Michel Makhlouf. Federated uncertainty-aware learningfordistributedhospitalehrdata.arXivpreprintarXiv:1910.12191, 2019.

[195] Pulkit Sharma, Farah E Shamout, and David A Clifton. Preserving patient privacy while training a predictive model of in-hospital mortality. arXiv preprint arXiv:1912.00354, 2019.

[196] Theodora S Brisimi, Ruidi Chen, Theofanie Mela, Alex Olshevsky, Ioannis Ch Paschalidis, and Wei Shi. Federated learning of predictive models from federated electronic health records. International journal of medical informatics, 112:59–67, 2018.

[197] Sawsan AbdulRahman, Hanine Tout, Hakima Ould-Slimane, Azzam Mourad, Chamseddine Talhi, and Mohsen Guizani. A survey on federated learning: The journey from centralized to distributed on-site learning and beyond. IEEE Internet of Things Journal, 8(7):5476–5497, 2020.

[198] Stephen R Pfohl, Andrew M Dai, and Katherine Heller. Federated and differentially private learning for electronic health records. arXiv preprint arXiv:1911.05861, 2019.

[199] Santiago Silva, Boris A Gutman, Eduardo Romero, Paul M Thompson, Andre Altmann, and Marco Lorenzi. Federated learning in distributed medical databases: Meta-analysis of large-scale subcortical brain data. In 2019 IEEE 16th international symposium on biomedical imaging (ISBI 2019), pages 270–274. IEEE, 2019.

[200] H Brendan McMahan, Daniel Ramage, Kunal Talwar, and Li Zhang. Learning differentially private recurrent language models. arXiv preprint arXiv:1710.06963, 2017.

[201] Keith Bonawitz, Hubert Eichner, Wolfgang Grieskamp, Dzmitry Huba, Alex Ingerman, Vladimir Ivanov, Chloe Kiddon, Jakub Konečný, Stefano Mazzocchi, Brendan McMahan, et al. Towards federated

learning at scale: System design. Proceedings of machine learning and systems, 1:374–388, 2019.

[202] CH Van Berkel. Multi-core for mobile phones. In 2009 Design, Automation & Test in Europe Conference & Exhibition, pages 1260–1265. IEEE, 2009.

[203] Jiawen Kang, Zehui Xiong, Dusit Niyato, Yuze Zou, Yang Zhang, and Mohsen Guizani. Reliable federated learning for mobile networks. IEEE Wireless Communications, 27(2):72–80, 2020.

[204] Angelo Silverio, Pierpaolo Cavallo, Roberta De Rosa, and Gennaro Galasso. Big health data and cardiovascular diseases: a challenge for research, an opportunity for clinical care. Frontiers in medicine, 6:36, 2019.

[205] Sricharan Donkada, Seyedamin Pouriyeh, Reza M Parizi, Meng Han, Nasrin Dehbozorgi, Nazmus Sakib, and Quan Z Sheng. Uncovering promises and challenges of federated learning to detect cardiovascular diseases: A scoping literature review. arXiv preprint arXiv:2308.13714, 2023.

[206] Dusit Thanapatay, Chaiwat Suwansaroj, and Chusak Thanawattano. Ecg beat classification method for ecg printout with principle components analysis and support vector machines. In 2010 International Conference on Electronics and Information Engineering, volume 1, pages V1–72. IEEE, 2010.

[207] Sricharan Donkada, Seyedamin Pouriyeh, Reza M Parizi, Meng Han, Nasrin Dehbozorgi, Nazmus Sakib, and Quan Z Sheng. Uncovering promises and challenges of federated learning to detect cardiovascular diseases: A scoping literature review. arXiv preprint arXiv:2308.13714, 2023.

[208] Jing Zhang, Deng Liang, Aiping Liu, Min Gao, Xiang Chen, Xu Zhang, and Xun Chen. Mlbf-net: A multi-lead-branch fusion network for multi-class arrhythmia classification using 12-lead ecg. IEEE journal of translational engineering in health and medicine, 9:1–11, 2021.

[209] Daoqin Lin, Yuchun Guo, Huan Sun, and Yishuai Chen. Fedcluster: A federated learning framework for cross-device private ecg classification. In IEEE INFOCOM 2022-IEEE Conference on Computer Communications Workshops (INFOCOM WKSHPS), pages 1–6. IEEE, 2022.

[210] A Rajkumar, M Ganesan, and R Lavanya. Arrhythmia classification on ecg using deep learning. In 2019 5th international conference on advanced computing & communication systems (ICACCS), pages 365–369. IEEE, 2019.

[211] Pranab Samanta, Akanksha Pathak, Kayapanda Mandana, and Goutam Saha. Classification of coronary artery diseased and normal subjects using multi-channel phonocardiogram signal. Biocybernetics and Biomedical Engineering, 39(2):426–443, 2019.

[212] Joseph Burns, Madhusudan Ganigara, and Arushi Dhar. Application of intelligent phonocardiography in the detection of congenital heart disease in pediatric patients: a narrative review. Progress in Pediatric Cardiology, 64:101455, 2022.

[213] Sumair Aziz, Muhammad Umar Khan, Majed Alhaisoni, Tallha Akram, and Muhammad Altaf. Phono- cardiogram signal processing for automatic diagnosis of congenital heart disorders through fusion of temporal and cepstral features. Sensors, 20(13):3790, 2020.

[214] Amirata Ghorbani, David Ouyang, Abubakar Abid, Bryan He, Jonathan H Chen, Robert A Harrington, David H Liang, Euan A Ashley, and James Y Zou. Deep learning interpretation of echocardiograms. NPJ digital medicine, 3(1):10, 2020.

[215] Yogen Singh. Echocardiographic evaluation of hemodynamics in neonates and children. Frontiers in Pediatrics, 5:201, 2017.

[216] Jia Zhou, Meng Du, Shuai Chang, and Zhiyi Chen. Artificial intelligence in echocardiography: detection, functional evaluation, and disease diagnosis. Cardiovascular ultrasound, 19(1):1–11, 2021.

[217] Shinichi Goto, Divyarajsinhji Solanki, Jenine E John, Ryuichiro Yagi, Max Homilius, Genki Ichihara, Yoshinori Katsumata, Hanna K Gaggin, Yuji Itabashi, Calum A MacRae, et al. Multinational federated learning approach to train ecg and echocardiogram models for hypertrophic cardiomyopathy detection. Circulation, 146(10):755–769, 2022.

[218] Bram Ruijsink, Esther Puyol-Anton, Ilkay Oksuz, Matthew Sinclair, Wenjia Bai, Julia A Schnabel, Reza Razavi, and Andrew P King. Fully automated, quality-controlled cardiac analysis from cmr: validation and large-scale application to characterize cardiac function. Cardiovascular Imaging, 13(3):684–695, 2020.

[219] Bram Ruijsink, Esther Puyol-Anton, Ilkay Oksuz, Matthew Sinclair, Wenjia Bai, Julia A Schnabel, Reza Razavi, and Andrew P King. Fully automated, quality-controlled cardiac analysis from cmr: validation and large-scale application to characterize cardiac function. Cardiovascular Imaging, 13(3):684–695, 2020.

[220] Justus Wolff, Julian Matschinske, Dietrich Baumgart, Anne Pytlik, Andreas Keck, Arunakiry Natarajan, Claudio E von Schacky, Josch K Pauling, and Jan Baumbach. Federated machine learning for a

facilitated implementation of artificial intelligence in healthcare–a proof of concept study for the prediction of coronary artery calcification scores. Journal of Integrative Bioinformatics, 19(4):20220032, 2022.

[221] Hangyu Zhu, Jinjin Xu, Shiqing Liu, and Yaochu Jin. Federated learning on non-iid data: A survey. Neurocomputing, 465:371–390, 2021.

[222] Xiang Li, Kaixuan Huang, Wenhao Yang, Shusen Wang, and Zhihua Zhang. On the convergence of fedavg on non-iid data. arXiv preprint arXiv:1907.02189, 2019.

[223] Chen Wang, Xinkui Wu, Gaoyang Liu, Tianping Deng, Kai Peng, and Shaohua Wan. Safeguarding cross- silo federated learning with local differential privacy. Digital Communications and Networks, 8(4):446–454, 2022.

[224] Felix Sattler, Simon Wiedemann, Klaus-Robert Mu¨ller, and Wojciech Samek. Robust and communication-efficient federated learning from non-iid data. IEEE transactions on neural networks and learning systems, 31(9):3400–3413, 2019.

[225] Tuo Zhang, Lei Gao, Chaoyang He, Mi Zhang, Bhaskar Krishnamachari, and A Salman Avestimehr. Federated learning for the internet of things: Applications, challenges, and opportunities. IEEE Internet of Things Magazine, 5(1):24–29, 2022.

3

Cybersecurity Kill Chain in Outer Space and Cyberspace Security

A. Geetha, V. Keerthika, and D. M. Deepak Raj

Department of Computer Science and Engineering, Alliance University, India

Email: geetha.a@alliance.edu.in; keerthika.v@alliance.edu.in; deepak.raj@alliance.edu.in

Abstract

The convergence of cyberspace and space presents new risks and vulnerabilities as technological development ushers in a period in which humankind is more dependent on infrastructure in space. In light of space activities and the necessity of safeguarding space-based systems, this book chapter investigates the application of the Cybersecurity Kill Chain framework. The Cybersecurity Kill Chain, which was originally created to counteract cyber threats on Earth, has been modified to take into account the particulars of environments in orbit.

The first section of the study outlines the various stages of the Space Cybersecurity Kill Chain, which include installation, weaponization, delivery, exploitation, reconnaissance, command and control, and actions on targets. Every stage is examined within the framework of space operations, taking into account the complexities of telemetry, satellite communication, and control systems.

The study also looks into how the threat environment in space is changing, with a focus on the increasing possibility of cyber–physical attacks against satellites, space-based assets, and ground-based control infrastructure. The conversation touches on how cybersecurity in space and on Earth are interdependent, emphasizing the ripple effects of a breach in one area.

The chapter suggests an extensive set of countermeasures and best practices to strengthen the security posture in cyberspace and outer space. Anomaly detection system integration, encryption of space communications, the creation of international partnerships to combat transboundary threats, and the creation of space-centric cybersecurity standards are a few of these.

The study ends with a recommendation for proactive and cooperative measures to protect space assets from new cyber threats by space-faring countries, regulatory agencies, and the commercial sector. Through the customization and execution of the Cybersecurity Kill Chain framework within the distinct domain of space exploration, interested parties can augment their capacity to identify, avert, and alleviate cyber hazards, guaranteeing the enduring viability and safety of space operations.

3.1 Introduction

In the contemporary world, technology has improved very fast as everyone wants to incorporate technology into their business, projects, and ideas. Most organizations need to participate in standards development activities toward cybersecurity and cyberspace. Cyberspace is a complex environment comprising interactions between people, software, and services, supported by information and communication technology (ICT) devices and networks. Cybersecurity awareness becomes a prime concern in today's network world.

Many of the organizations are not seriously involved in cyberspace or cybersecurity. They may feel that it is optional to influence the outcome, organizations can think of participating, depending upon the importance of the standard to the organization. They should consider becoming members of the entity developing the standards toward cybersecurity.

The cyber kill chain [4] adapts the military's kill chain, a step-by-step approach that identifies and stops enemy activity. The cyber kill chain [4] outlines the various stages of several common cyber-attacks, where the information security team can prevent, detect, or intercept attackers. It ultimately provides information about investigation, command and control, action, weaponization [4], delivery, exploitation, and installation. This model can be helpful for companies to be aware and proactive in preventing such attacks in the future to avoid massive loss of resources and data.

Even if the organizations have not taken a more active role in their cybersecurity standard development, they need to predict its consequences by following the development of a new standard. The content and impact of a standard still in development can be influenced by public reviewers who examine standard draughts and provide comments. Organizations should consider joining the body to create the standards if it is exceptionally significant. The driver's position may be appropriate when the organization's commitment to a new standard is crucial. Driving the creation of a standard may take significant resources; it is possible that producing the standard is part of the organization's charter or goal.

Organizations should consider taking part in the maintenance of current standards in addition to helping to create new ones; most standards are regularly reviewed and updated. Indicator SDOs often have formal maintenance programs to help guarantee that their standards do not become outdated because of changes in related standards, technological advancements, or other factors.

The concept of the Cybersecurity Kill Chain has been adapted and extended to address the unique challenges and threats in outer space and cyberspace security. It provides a structured framework for understanding and defending against attacks on critical space-based assets and their associated cyber infrastructure.

3.2 Framework Environment

A cyber kill chain framework can help organize a better understanding of relevant threats and improve incident management and response. The cybersecurity kill chain is a framework used to describe the various stages a cyberattack goes through, from the initial reconnaissance to the final objective of the attacker. It takes on a slightly different context when applied to outer space and cyberspace security, but the principles remain largely the same. Figure 3.1 depicts the Framework for Cybersecurity Kill Chain.

3.2.1 Security concerns in outer space and cyberspace

Security concerns in outer space and cyberspace are growing due to the increasing reliance on satellite technology and the interconnectedness of our digital world. These concerns encompass a range of issues.

3.2.1.1 Reconnaissance (gathering intelligence)

A malicious actor finds a target and investigates network flaws and vulnerabilities that can be used against it during reconnaissance. During this process, the attacker might harvest login credentials or obtain additional data, including user IDs, email addresses, physical locations, software, and operating system details, all of which could be helpful in phishing attacks. Generally speaking, an attack will be more sophisticated and convincing and

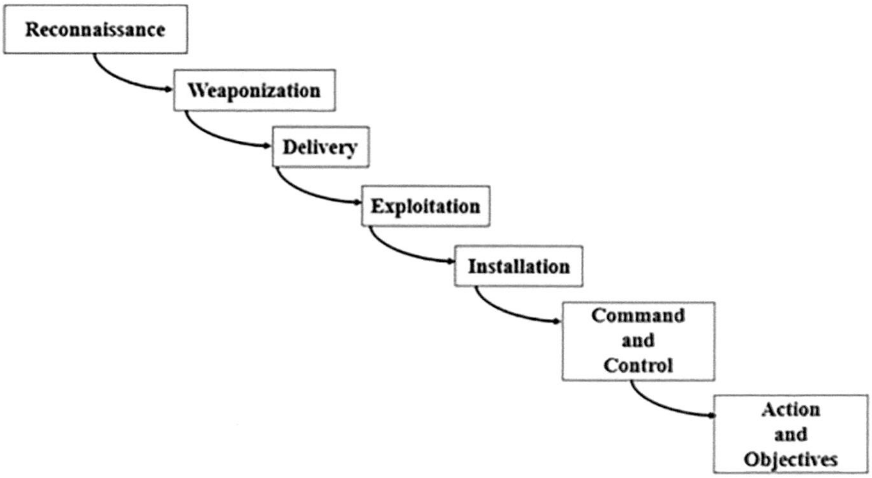

Figure 3.1 Framework for Cybersecurity Kill Chain.

more likely to succeed if the attacker obtains more information during the reconnaissance phase. In outer space security, this stage involves collecting information about satellite orbits, communication protocols, and potential vulnerabilities in space-based systems. In cyberspace security, this is about collecting information on potential targets, including vulnerabilities, system configurations, and user behaviors.

3.2.1.2 Weaponization (creating malicious tools)

In the weaponization stage, the attacker develops an attack vector – a virus, worm, ransomware, or remote access malware – that can exploit a known vulnerability. In this stage, the attacker might also install back doors, allowing them to enter the system even if network administrators find and shut down their initial entry point.

In this context, weaponization could involve developing or acquiring technology capable of disrupting satellite communications or disabling satellites. Weaponization [4] involves creating malware [8] exploits or other cyber tools to exploit vulnerabilities identified during reconnaissance [13].

3.2.1.3 Delivery (introducing malware or payload)

Payloads or malicious technology might be deployed via satellites or anti-satellite weapons. Malware [8] or other cyber threats are delivered through various means, such as phishing emails, infected websites, or USB drives. For example, the attacker might provide malicious links or email attachments to encourage user interaction and forward the plot. This activity could be coupled with social engineering strategies to boost the campaign's efficacy.

3.2.1.4 Exploitation (taking advantage of vulnerabilities)

The malicious code is executed within the victim's system in this phase. An attacker might, for instance, be unable to access a company's databases from the outside. However, once inside, they could discover that a database is running an outdated version and is vulnerable to a well-known flaw. Examples are local job scheduling and dynamic data exchange.

3.2.1.5 Installation (establishing a foothold)

The virus or other attack vector will be put on the victim's machine right after the exploitation phase. The threat actor has now gained access to the system and can take control, marking a significant turning point in the assault life cycle.

For outer space, this stage may involve establishing control over compromised satellites or space assets. In cyberspace, attackers install a persistent backdoor or establish control over compromised systems.

3.2.1.6 Command and control (managing the attack)

Through command and control, an attacker can remotely take control of a device or identity on the target network using malware. At this point, the attacker may also migrate laterally across the network to increase their access and set up other entry points in the future. Attackers will seek to maintain control over compromised space assets. In cyberspace, this involves controlling the compromised systems, often through remote servers.

3.2.1.7 Actions on objectives (achieving goals)

The attacker moves forward with their planned objectives, which could involve exfiltration, encryption, data theft, or destruction. In outer space, the objective may be to disrupt satellite communications, steal sensitive data, or deny services. The ultimate goal could be data theft, financial fraud, espionage, or malicious intent. Organizations should, therefore, reduce the danger of an attack and the expense of resolving an event by taking action to detect and neutralize threats as early in the life cycle as feasible.

3.3 Architecture Diagram

Concerning network security, the "kill chain" framework can be applied to describe the stages that an attacker might go through to successfully compromise a network or system. It helps organizations understand and defend

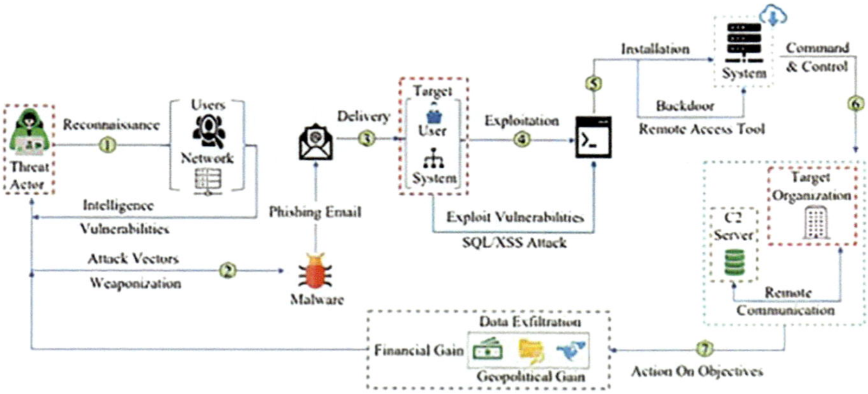

Figure 3.2 Architecture of the Cybersecurity Kill Chain [6].

against cyber threats by identifying vulnerabilities and implementing countermeasures at each stage [6]. The kill chain stages in the network architecture context are sketched in Figure 3.2.

Threat actor: A "threat actor" is a term commonly used in cybersecurity to refer to any individual, group, organization, or entity that poses a potential threat to a computer system, network, or digital environment. Threat actors are responsible for carrying out various malicious activities, including cyber-attacks [3], data breaches, and other unauthorized actions that compromise digital assets' confidentiality, integrity, or availability.

Target user: In the field of cybersecurity, a "target user" refers to an individual, group, or entity who is focused on potential cyber threats, attacks, or security measures [3] and needs protection [4] from various online risks.

Understanding and effectively securing target users is essential for maintaining digital information and systems' confidentiality, integrity, and availability.

Attack vectors [1]: A specific path or methods that threat actors use to exploit system, network, or application vulnerabilities to carry out a cyber-attack. Attack vectors [1] outline how an attacker gains unauthorized access, spreads malware [8], or conducts other malicious activities. These vectors can be diverse [6] and depend on the target's architecture, configuration, and security weaknesses.

Phishing: Attackers use deceptive emails or messages to trick users into revealing sensitive information, clicking on malicious links, or downloading infected attachments.

This process primarily aims to prevent phishing attacks at the early stage of the kill chain.

Vulnerabilities: Attackers identify and exploit vulnerabilities in software or systems to gain unauthorized access [5]. This could involve exploiting known software vulnerabilities or zero-day vulnerabilities.

SQL injection/cross-site scripting (XSS): Cross-site scripting is abbreviated as XSS. In this attack, attackers manipulate input fields of a website or application to execute unintended SQL commands, potentially gaining access to databases or data. Attackers inject malicious scripts into trusted websites, which then execute in the users' browsers, potentially stealing information or executing unauthorized actions [7]. Publicly available or visible profile pages are targets for persistent XSS. The profile boxes, injected with malicious scripts in their browser, will execute the code automatically when the user opens their profile. Reflective XSS is obscured URLs sent to the targeted user.

Exploits: An exploit is a code or technique that leverages a vulnerability to achieve a specific goal. Exploits can execute arbitrary code, gain unauthorized access, escalate privileges, or carry out other malicious activities.

Data exfiltration: Unauthorized transfer or removal of sensitive, valuable data from a computer system, network, or organization to an external location controlled by malicious actors, also known as data theft, involves stealing and transferring sensitive information. Outbound email systems of organizations, such as calendars, databases, images, and planning documents, tend to be attacked by cyber attackers.

Remote access control refers to the ability to manage and control computer systems, networks, and devices from a remote location. This capability is essential for IT administrators, support teams, and authorized users to maintain, troubleshoot, and manage systems without being physically present at the location of the devices they are working on. However, remote access control also presents security challenges, as unauthorized access could lead to data breaches, system compromise, and other cybersecurity risks [12].

Financial gain is acquiring money, profits, or monetary benefits through various legitimate or illicit means. In cybersecurity and cybercrime, financial gain can involve legal and illegal activities that result in monetary rewards. The criminal will then take over communications with customers or vendors while doing business illegally.

Backdoors: A "backdoor" in cybersecurity refers to a hidden or unauthorized method of accessing a computer system, application, or network. It is intentionally created or exploited by attackers to bypass security mechanisms. The most effective backdoors are cleverly hidden, providing unauthorized users [14] with an entry point that is difficult to detect.

A "target organization" relates to cybersecurity and refers to a specific company, entity, institution, or government agency that is the subject of a potential cyber-attack, data breach, or other cyber threats. Target organizations are those adversaries identified as potential victims and aim to compromise, exploit, or breach for various reasons, including financial gain, espionage, disruption, or ideological motivations.

3.3.1 Cybersecurity Kill Chain process

The "Cybersecurity Kill Chain" is a framework that describes the various stages of a cyber-attack, from the initial reconnaissance to the final data exfiltration or impact. It provides a structured approach for understanding and analyzing the steps attackers go through to achieve their objectives. Organizations can develop effective strategies to prevent, detect, and respond to cyber threats by breaking down the attack process into distinct phases. This framework helps organizations better understand an attacker's tactics and techniques, enabling them to implement targeted security measures at each stage.

3.3.2 Principle

The "kill chain" is a concept used in various fields, including military and cybersecurity, to describe a series of stages or steps an adversary goes through to achieve a specific objective. In cybersecurity, the kill chain principle refers to a structured framework that breaks down the different stages of a cyber-attack, helping organizations understand the attacker's tactics and enabling them to develop strategies to prevent, detect, and respond to threats effectively. The kill chain principle aims to disrupt or mitigate attacks at various points along the attack life cycle.

3.3.3 The basic principles of the Cybersecurity Kill Chain

Understanding the attack life cycle: The kill chain framework helps organizations comprehend the steps that attackers take, from the initial reconnaissance [13] to the final data exfiltration. This understanding allows defenders to anticipate attacker behavior.

Proactive defense: By identifying the stages of an attack, organizations can take proactive measures to prevent or disrupt attacks at different points. This involves implementing security controls and best practices that correspond to each stage.

Layered defense: Instead of relying solely on a single security measure, the kill chain principle emphasizes the importance of a layered defense. Different security controls are deployed at different stages to create multiple barriers for attackers.

Continuous monitoring: Effective cybersecurity requires constant monitoring and analysis of network and system activities. This helps identify anomalies, detect ongoing attacks, and respond promptly.

Threat intelligence: Understanding threat actors' tactics, techniques, and procedures (TTPs) is crucial. Threat intelligence provides insights into emerging threats, enabling organizations to adapt their defenses.

Incident response: Organizations need well-defined incident response plans that outline how to react to potential attacks at each stage of the kill chain. A coordinated response is essential for minimizing damage.

Adaptability: The kill chain principle acknowledges that attackers' tactics evolve. Organizations must continually adapt their defenses and strategies to counter new threats.

Collaboration: Sharing threat intelligence and collaborating with other organizations, industry groups [13], and government agencies can enhance collective defense against cyber threats.

Forensic analysis: After an attack occurs, detailed forensic analysis can provide insights into the attack's origin, methods, and potential weaknesses in the organization's defenses.

Continuous improvement: Regularly reviewing and refining the organization's cybersecurity strategy based on lessons learned from previous incidents and threat intelligence is vital for maintaining a solid security posture. Overall, the kill chain principle provides a systematic approach for organizations to defend against cyber threats. By breaking down attacks into discrete stages and applying appropriate security measures at each stage, organizations can reduce the likelihood of successful breaches and mitigate the impact of potential security incidents.

3.4 Challenges of Kill Chain in Outer Space and Cyberspace Security

Implementing the kill chain concept in outer space and cyberspace security comes with several challenges and considerations described in Table 3.1.

3.5 Applications of Kill Chain in Outer Space and Cyberspace Security

When applied to outer space and cyberspace security, the kill chain concept offers several advantages for organizations and nations seeking to protect their assets and interests; the kill chain model [8] is a valuable tool for enhancing security and resilience in outer space and cyberspace. It helps organizations and governments stay ahead of evolving threats and ensures a proactive, structured approach to security.

Let us examine a hypothetical case study of the kill chain concept applied to outer space and cyberspace security to illustrate its use and effectiveness in addressing security threats in these domains.

Cyberspace security continues to evolve to address emerging threats and technological advancements. Here are some trends in how the kill chain concept is being applied in these domains:

(i) Integration of cyber and space security:
The boundaries between cyber and space security are becoming blurred as satellites and ground-based control systems rely heavily on interconnected computer networks. The kill chain concept is increasingly applied to cyber and space security aspects to create a unified defense strategy.

(ii) Automation and AI:
Automation and artificial intelligence (AI) are growing to enhance threat detection and response at various stages of the kill chain. AI-driven algorithms can analyze vast datasets in real time to identify anomalies and potential threats in both cyber and space environments.

(iii) Threat intelligence sharing:
Collaborative information sharing and threat intelligence [8] sharing among governments, organizations, and international bodies is rising. This trend helps identify threats early in the kill chain and facilitates proactive defenses.

(iv) Zero trust architecture:
A zero trust security model is gaining prominence in cyberspace and outer space. It involves verifying the identity and security of users and devices

Table 3.1 Difference between cyberspace and outer space.

Cyberspace	Outer space
Cyber threats constantly evolve, and attackers may not follow a linear kill chain model [8]. They adapt quickly, making anticipating and defending against their tactics challenging.	Threats in outer space, such as anti-satellite weapons, can be highly complex and may not fit neatly into the traditional kill chain stages.
Attribution difficulties lead to potential misidentification of attackers and complicated responses.	Identifying the source of a threat, such as a satellite jamming signal, can be similarly complex.
Resource constraints: Implementing comprehensive cybersecurity measures requires significant resources, which may challenge smaller organizations or nations.	Building and maintaining secure satellite systems can be costly, limiting access to some entities.
Balancing security with individual privacy rights can be challenging, especially when monitoring user behavior during threat detection.	Space-based surveillance and tracking systems raise concerns about potential privacy violations and international norms.
Achieving international cooperation and agreement on cybersecurity standards and norms can take time due to differing national interests.	Forming agreements and norms related to space security is a complex diplomatic challenge, as countries may have conflicting interests.
Focusing too much on preventing attacks may lead to neglecting incident response and recovery capabilities.	A strong emphasis on preventing attacks may not address vulnerabilities in existing satellite systems.
Human error and insider threats are significant challenges, as employees may inadvertently or intentionally compromise security.	Insider threats can also pose risks, as personnel with access to space systems may misuse their privileges.
The legal landscape for cyber operations is complex and varies from country to country, making it challenging to establish consistent rules.	The legal framework for outer space activities, including security measures, is still evolving, leading to uncertainties.

at every stage, reducing the attack surface, and minimizing the impact of breaches.

(v) Multilayered defense:

Organizations and governments are adopting multilayered defense strategies, combining traditional security measures like firewalls and intrusion detection [1] systems with advanced threat hunting and response capabilities to address attacks at different stages of the kill chain.

(vi) Focus on supply chain security:
There is an increasing focus on securing the supply chain, both in terms of the hardware and software components used in satellites and critical cyber infrastructure. Attackers can exploit weak links in the supply chain.

(vii) Regulatory frameworks:
Governments and international organizations are working on establishing regulatory frameworks and behavior norms in outer space and cyberspace. These frameworks aim to create rules of engagement and promote responsible conduct, aligning with the principles of the kill chain concept.

(viii) Threat emulation and red teaming:
Organizations are conducting more advanced threat emulation and red teaming exercises to simulate real-world attack scenarios and assess their readiness to respond at different stages of the kill chain [2].

(ix) Cloud and IoT security:
As cloud computing and the Internet of Things (IoT) expand, security challenges related to these technologies are incorporated into the kill chain concept. Special attention is given to securing cloud-based infrastructure and IoT devices.

(x) Resilience and continuity planning:
Organizations are focusing on building resilience and continuity planning into their security strategies. This involves preparing for the worst-case scenario, such as a complete disruption of satellite communication or a significant cyber incident.

(xi) Quantum threat mitigation:
Researchers and security experts are exploring strategies to mitigate the potential threats of quantum computing to encryption in cyberspace and outer space. Post-quantum cryptography is a growing field of interest. Professionals in outer space and cyberspace security are receiving cross-domain training to understand the unique challenges and threats in each environment and develop integrated security strategies.

3.5.1 Application: securing a satellite communication network

Scenario: A government agency operates a satellite communication network critical for national security, providing secure communication links to military units and intelligence [8] agencies. This example deals with outer space security with the stages of the cybersecurity kill chain.

Reconnaissance: In outer space, the reconnaissance stage involves monitoring satellite orbits and radio frequency emissions to gather intelligence on the satellite network [8].

Weaponization: Adversaries develop ground-based anti-satellite weapons capable of disrupting satellite operations.

Delivery: These weapons are delivered through ground-based radar systems and missile launch platforms.

Exploitation: The adversaries use anti-satellite weapons to exploit vulnerabilities in the satellite network.

Installation: The successful attack allows adversaries to gain control of one of the satellites, establishing a foothold in space.

Command and control: The attackers establish command and control over the compromised satellite, enabling them to manipulate communication links.

Actions on objectives: The attackers disrupt communication between military units, creating confusion and potentially compromising national security.

Exfiltration: While not applicable in this context, attackers may attempt to intercept sensitive data transmitted through the compromised satellite.

Covering tracks: Attackers employ diversionary tactics, launching additional attacks on unrelated targets to divert attention from the satellite breach.

Post-attack: The attackers maintain control over the compromised satellite, enabling them to carry out further attacks or demand a ransom for restoring communication services.

3.6 Summary

The kill chain concept is adapting to the changing threat landscape and the increasing convergence of cyber and space security. The trends involve a more integrated and proactive approach, emphasizing information sharing, advanced technologies, regulatory frameworks, and resilience planning to secure critical infrastructure and national interests in both domains.

When applied to outer space and cyberspace security, the Cybersecurity Kill Chain is a framework used to understand and mitigate security threats in these domains. It breaks down the stages of a potential attack, allowing for proactive defense strategies. Outer space security protects satellite systems from reconnaissance, weaponization, and attacks [13]. Cyberspace security covers the stages from reconnaissance to actions on objectives, focusing on identifying, preventing, and responding to cyber threats. This concept emphasizes early threat detection, risk assessment, and collaborative international efforts to safeguard critical infrastructure and national interests in outer space and cyberspace.

References

[1] HosseiniNejad, HaddadPajouh, H., Dehghantanha, A., and Parizi, R.M. "A Cyber Kill Chain Based Analysis of Remote Access Trojans," in Handbook of Big Data and IoT Security, A. Dehghantanha and K.- K. R. Choo, Eds. Cham: Springer International Publishing, 2019, pp. 273–299. doi: 10.1007/978-3-030- 10543-3_12

[2] Adrian Duncan, Sadie Creese, and Michael Goldsmith. "A Combined Attack-Tree and Kill-Chain Approach to Designing Attack-Detection Strategies for Malicious Insiders in Cloud Computing," in 2019 International Conference on Cyber Security and Protection of Digital Services (Cyber Security), Oxford, United Kingdom, Jun. 2019, pp. 1–9. doi: 10.1109/CyberSecPODS.2019.8885401.

[3] Hoffmann, R. (2019). Markov Models of Cyber Kill Chains with Iterations. 1–6. 10.1109/ICMCIS.2019.8842810.

[4] Ira Winkler, Araceli Treu Gomes. Chapter 16 - Kill Chain Analysis, Editor(s): Ira Winkler, Araceli Treu Gomes, Advanced Persistent Security, Syngress,2017, Pages 179-184, ISBN 9780128093160, https://doi.org /10.1016/B978-0-12-809316-0.0 0016-6.

[5] Du, J., Zhang, X., Suo, G., Guo, R., and Lu, G. "A Method of Network Behavior Recognition and Attack Scenario Reconstruction for Attack Kill Chain," International Conference on Wireless Communication, Network and Multimedia Engineering (WCNME 2019), Guilin, China, 2019.

[6] Ju, Ankang and Guo, Yuanbo and Ye, Ziwei and Li, Tao and Ma, Jing. (2019). HeteMSD: A Big Data Analytics Framework for Targeted Cyber-Attacks Detection Using Heterogeneous Multisource Data. Security and Communication Networks. 2019. 1–9. 10.1155/2019/5483918.

[7] Jin, X. (2014). Code Injection Attacks on HTML5-based Mobile Apps: Characterization, Detection and Mitigation. CCS'14, Scottsdale, Arizona, USA. Copyright 2014 ACM 978-1-4503-2957-6/14/11 http://dx.doi.org/10.1145/2660267.2660275.

[8] Khan, M.S., Siddiqui, S., and Ferens, K. (2018). A Cognitive and Concurrent Cyber Kill Chain Model. In: Daimi, K. (eds) Computer

and Network Security Essentials. Springer, Cham. https://doi.org/10.1007/978-3-319-58424-9_34

[9] Manulis, M., Bridges, C.P., and Harrison, R. et al. Cyber security in New Space. Int. J. Inf. Secure. 20, 287–311 (2021). https://doi.org/10.1007/s10207-020-00503-w

[10] MuhammadMudassarYamin,MohibUllah,HabibUllah,Katt,B.,Hijji,M., and Muhammad, K. "Mapping Tools for Open Source Intelligence with Cyber Kill Chain for Adversarial Aware Security," Mathematics, vol. 10, no. 12, p. 2054, Jun. 2022, doi: 10.3390/math10122054

[11] Straub, J. (2020). Modeling Attack, Defense, Threat Trees, Cyber Kill Chain, ATT&CK, and STRIDE Frameworks as Blackboard Architecture Networks. 148–153. 10.1109/SmartCloud49737.2020.00035.

[12] Thuraisingham, B., Kantarcioglu, M., Hamlen, K., Khan, L., Finin, T., Joshi, A., Oates, T., and Bertino, E. (2016). A Data Driven Approach for the Science of Cyber Security: Challenges and Directions. 1–10. 10.1109/IRI.2016.10.

[13] Xiaojun Zhou, Zhen Xu, Wang, L., Chen, K., Chen, C., and Zhang, W. "Kill Chain for Industrial Control System," MATEC Web Conf., vol. 173, p. 01013, 2018, doi: 10.1051/matecconf/201817301013.

[14] Ahamed, N. N., and Karthikeyan, P. (2020). A reinforcement learning integrated in heuristic search method for self-driving vehicle using blockchain in supply chain management. International Journal of Intelligent Networks, 1, 92-101.

4

Ransomware Attacks in Cyber Space and Mitigation Strategies

K. Rajkumar[1], S. Karthikeyan[2], V. Kavitha[1], and U. Hariharan[3]

[1]Department of Data Science and Business System, School of Computing, SRM Institute of Science and Technology, Kattankulathur, Tamil Nadu, India
[2]Department of Computer Science and Engineering, Faculty of Engineering and Technology, Jain (Deemed-to-be University), India
[3]Department of Computer Science and Engineering, Apex Institute of Technology, Chandigarh University, India

Email: rajkumak5@srmist.edu.in; kavithv8@srmist.edu.in; link2karthikcse@gmail.com; hariharan.e11201@cumail.in

Abstract

Ransomware is used to block file access for the user or any institute, where access will be given by paying some ransom to the attackers. The ransomware attack scatters quickly and has severe effects, including severe data loss, destruction of data, reputation damage, and downtime for organizations. It is hazardous for hospitals, schools, and government organizations that rely on that information to function daily. Phishing emails, visiting corrupt websites, and remote desktop protocols are common attacks in which attackers trap users. Even though ransomware attacks are increasing, there are practices such as backing up data, updating software systems, antivirus, and firewall protection, email protection, and regular security testing to ensure your privileged accounts are strictly controlled. Intruders attack the system and ransom the money, and it is not guaranteed that the intruders will be given the encryption key after paying the ransom. Thus, the ransomware malware detection analysis and detection approach like the signature-based detection approach will be able to prevent and detect the intruders before/after entering the system. This study provides an ample list of attacks, such as crypto-ransomware and locker ransomware, that intruders can make and ways to prevent the system.

4.1 Introduction

The development of the internet as a metaphor and the expanding impact of telecommunication networks have completely changed how businesses operate. Various malicious software applications, including CryptoLocker, Locky, CryptoWall, KeyRanger, SamSam, TeslaCrypt, TorrentLocker, and others, are collectively called "ransomware." Unheard-of cybercriminal behaviors have emerged due to the phenomenal advancements in technology, digitalization, cloud computing, fog computing, edge computing, quantum computing, and the widespread deployment of linked things. From a personal and social perspective, the increased risks of a covert cyber-attack on government agencies, financial institutions, data centers, and vital infrastructures pose significant concerns. Malicious software, commonly known as malware, is used in sophisticated cyber-attacks, and cyber-espionage is the exfiltration of critical data, disruption, identity theft, and other political motives.

While the worldwide computer industry is making incredible strides, cybercrime has expanded internationally. According to the Kaspersky Security Bulletin, hackers attacked users in 2022 with 400,000 new malicious files daily, which is an increase of 5% from 2021. According to Kaspersky's Security Bulletin, the daily occurrence of ransomware has increased by 181% since 2021, affecting 9500 encrypted files daily [1]. Security experts from Kaspersky found a daily 10% increase in the percentage of malicious files aimed at the Android platform. Some of the recent ransomware attacks carried out in India are as follows [2]:

1. The WannaCry ransomware attack: This affected more than 2 lakh computer systems. This ransomware attack hit banks in India and a few enterprises in Tamil Nadu and Gujarat.

2. Mirai botnet malware attack: Another shocking ransomware attack in India was the Mirai botnet malware attack.

3. AIIMS attack: In 2023, hackers attacked AIIMS Delhi, causing server shutdowns and disrupting health services. Patient data was potentially compromised, highlighting the dangers of cyber-attacks on healthcare. This malicious attack emphasized the need for more robust cybersecurity measures to protect sensitive information.

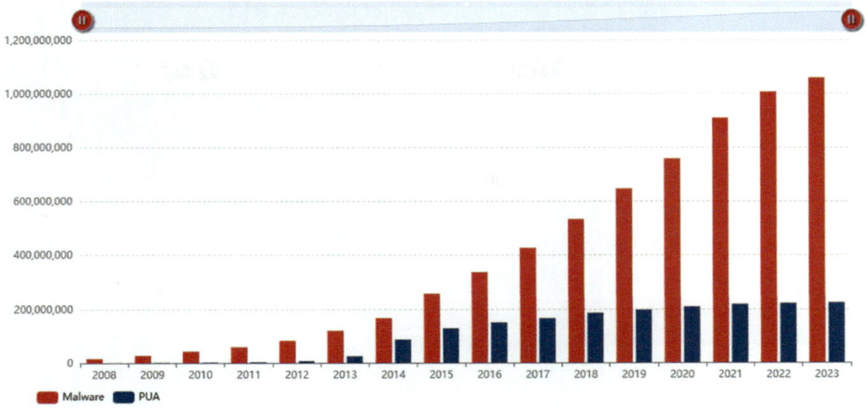

Figure 4.1 Total amount of malware and PUA.

4. Telangana and Andhra Pradesh power utility systems attack: Telangana and Andhra Pradesh, two southern states in India, got hit by a ransomware attack on their power utility systems 2019. The malicious virus took down all the servers until the glitch was resolved [10]. Since the computer systems of the two states were interlinked, the virus spread quickly, leading to a complete shutdown of all systems.

5. UHBVN ransomware attack: Uttar Haryana Bijli Vitran Nigam, a power company in Haryana, was attacked by hackers who gained access to the company's computer systems and stole customers' billing data. The attackers, in return, demanded a hefty ransom amount of Rs.1 crore or $10 million in exchange for returning the stolen data.

Furthermore, every day, the AV-TEST Institute records about 450,000 new harmful programs (malware) and potentially unwanted applications (PUAs) [3][4]. Figure 4.1 shows the estimated total number of PUAs and malware for 2022. For the 2022 SpyCloud Ransomware Defence Report, over 300 IT security professionals working for US, UK, and Canadian enterprises were surveyed [8][9], which asked at least 500 workers how seriously they took the danger of ransomware. In 2022, 90% of organizations were hit by ransomware, according to the poll [17], which is a substantial rise from 2021, when the figure was 72.5%. According to Symantec, more than 50% of new viruses are variations of existing malware [7].

Cybercriminals use both known and undiscovered weaknesses to build effective viruses. The size of malware attacks that may be carried out on these devices is easily imaginable, given the widespread deployment of so many

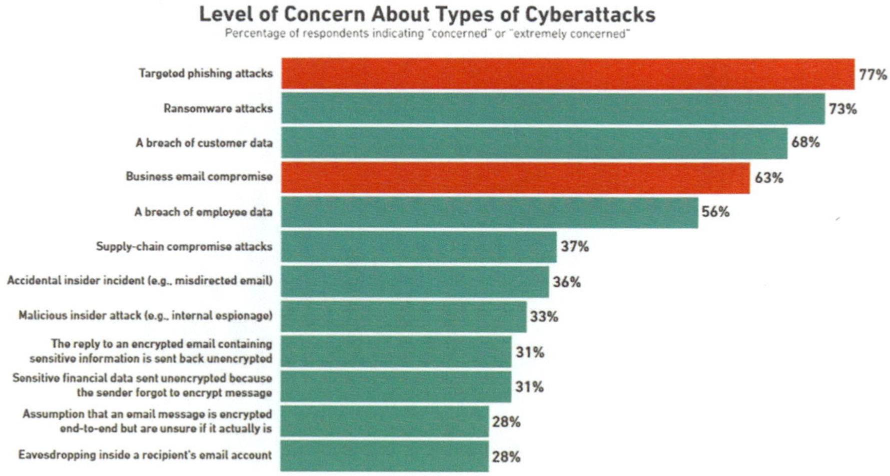

Figure 4.2 Types of malware distribution.

interconnected items [34]. This is a serious issue that needs to be resolved. Governments and law enforcement may continue to concentrate on ransomware shortly because of the risks presented by infrastructure assaults and data breaches. However, these patterns suggest that organizations are keeping an eye on the broader picture even while other assault types continue to expand and become more severe. Ransomware assaults are compared to some of the other attacks [13]. Figure 4.2 shows malware distribution from January 2022 to June 2022.

The chapter is organized as follows. Section 4.2 gives the different types of ransomware. The way people get ransomware and examples of ransomware attacks are discussed. Section 4.3 discusses the methods to handle ransomware, such as crypto-ransomware and locker ransomware. Section 4.4 discusses research on ransomware malware detection analysis (static, dynamic, and hybrid) and detection approaches like signature-based detection approach, behavioral-based detection approach, and hybrid detection approach. Section 4.5 compares and highlights the most common related work for malware detection and analysis. Some discussion on ransomware prevention and removal follows Section 6. Finally, Section 7 presents our conclusion.

4.2 Different Types of Ransomwares

There are four primary categories of ransomware: locked and crypto-ransomware, scareware, and doxware. Both ransomware programs often

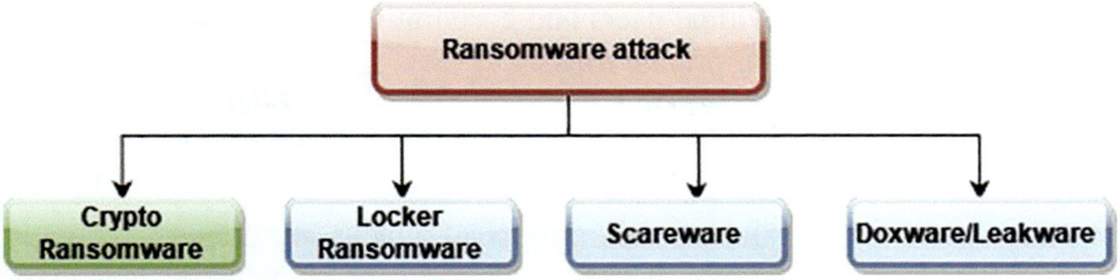

Figure 4.3 Taxonomy of ransomware attack.

begin with an email attachment or a web link; when the victim opens the file or clicks on the link, the ransomware begins to assault the victim's machine by exploiting known operating system faults. Figure 4.3 depicts the ransomware attack taxonomy.

Crypto-ransomware: This sort of ransomware, which is the most common, encrypts the files, data, or information on the victim's device. The victim may typically log into the system, access, and even use the data. However, because the data is encrypted, they could not access it. Additionally, victims of crypto-ransomware are required to pay. If the user were late, all encrypted data would be permanently lost.

Lockers: These types of ransomware prevent people from accessing their computers. User interaction with screens that show the ransom demand or the lock screen is typically restricted. The attacker could use the mouse and keyboard to make a partial payment. Lockers often do not delete the data; they only prevent people from accessing it.

Scareware: Typically, scareware poses problems on the infected device. The user often receives notification of a discovered virus or malware in the form of a message on their screen. You can click the alert during the panic and let the malware loose. By presenting an unsettling message, scareware typically seeks to freak off users to deceive them into downloading malware. Attackers frequently entice victims to respond quickly without giving them much time to consider or analyze their options by using prompts that appear authoritative and legitimate. The instructions could appear as a pop-up window, a menacing message, or a phony button with alarming wording like "Your PC is slow, Speed up Now," or "Attackers can see your IP, Protect it now," for instance. By accepting the bait, users allow the ransomware to infiltrate their computers, lock them out, and encrypt their data.

Doxware/Leakware: Doxware frequently uses threats of disclosing private information online to trick a person or a business into paying a ransom.

Doxware can target individual users and corporations, threatening to release proprietary information or private files.

4.2.1 Ways people get ransomware

Visiting corrupted websites: Quick download attacks are defined explicitly as unauthorized installations of malicious software on your devices. This also holds for unintentional computer system downloads of files or software packages. Even 100% authentic websites are compelled to spread this risk by these attacks because they are invisible online. Variants: The two primary Drive-by-Download assault types are as follows: non-malicious potentially unwanted programs or applications (PUPs/PUAs). Malware-loaded attacks include the following: downloading malicious attachments or file extensions, and exploiting vulnerabilities in networks and systems.

Phishing emails: When someone sends a phony email that appears to be from a reliable source, it is known as a phishing attack. Email is typically used for this. Either malware will be installed on the victim's machine, or personal data like credit card numbers and login credentials will be taken. Everyone should educate themselves about phishing, a prevalent cyber-attack, to defend themselves against it.

Microsoft's proprietary remote desktop protocol (RDP) enables remote connections between devices running any operating system (OS). RDP enables IT managers to diagnose staff members' problems remotely while granting them access to company resources. Even though they are proprietary, some RDP standards are open, allowing anyone to use them to enhance the protocol's features and, if necessary, adapt them to organizational needs.

4.2.2 Examples of ransomware

CryptoLocker: One of the first ransomware variants, CryptoLocker, encrypts its victims' data and demands payment in Bitcoins or pre-paid gift cards for the private key needed to decrypt it. If the deadline is not reached, the attackers threaten to permanently restrict access to the data by wiping the private key.

NotPetya: This virus directly targets the computer's boot record and encrypts the NTFS file system. As a result, until the ransom is paid, the system cannot boot into the operating system. Some observers have categorized it as a nation-state's "act of war" against Ukraine rather than the aim for financial gain of hackers.

Bad Rabbit: Bad Rabbit encrypts the user's file tables and demands Bitcoins to decrypt them. Governmental organizations like the Ministry of Infrastructure of Ukraine were primarily affected. Several of the code's strings contained names of characters from the well-known television series Game of Thrones.

Locky: In 2016, a well-organized cyber gang carried out the first ransomware attack. Locky encrypted files in over 160 distinct file types and spread via phony emails with malware attachments. The malware was placed on the systems of users who fell for the email hoax.

Ryuk: The recovery feature of Windows operating systems was disabled by the encryption Trojan Ryuk, which began to propagate in August 2018. Without an external backup, it was impossible to recover the encrypted data.

WannaCry: WannaCry, a ransomware infection purportedly created by the North Korean Lazarus Group, blended bespoke code with exploit code taken from the US government. According to estimates, the worm was used in a worldwide attack in May 2017 that infected 200,000 machines in three days. The virus may infect new victims by taking advantage of a Windows PC flaw, allowing it to propagate quickly across the Internet.

Revil: The ransomware-as-a-service (RaaS) organization Revil has extorted substantial sums of money from businesses worldwide during the past year. The Resident Evil movie franchise inspired the name Ransomware Evil.

4.3 The General Method for handling a Ransomware Attack

Step 1: Getting ready. The most excellent way to strengthen defenses is to patch systems aggressively because malware frequently accesses systems through known flaws. By removing weaknesses, you might prevent the malware from accessing any of your systems in the first place.

Step 2: Detection. If enterprises find the malware quickly after an assault, they can reduce harm. Getting signatures and IOCs into an IDS or other network device is a good safeguard against initial exploitation and infection.

Step 3: Containment. There are ways to stop the ransomware from spreading to other devices after it has already infected one, protecting network files in the process.

Figure 4.4 Working principle for crypto-ransomware.

Step 4: Eradication. Eliminating the ransomware problem from the network comes after it has been located and contained. Machine replacement is typically advised rather than cleaning. It can be challenging to determine whether residual files are present on the system and capable of reinfecting devices, as with all malware.

Step 5: Recovery. Any ransomware incident may be transformed into a non-issue if there are solid, verified backups by simply replacing or cleaning computers and recovering from backups.

Handling the crypto-ransomware: In this crypto-ransomware case, the malware encrypts some crucial user files once it activates. In this instance, ransomware selects a few critical files based on file extensions rather than attacking the entire victim's hard drive. A nearly impossible 24-bit encryption technique is typically used in this attack. Instead of opening the attachment or clicking the link, exploit kits attack victims through corrupted, missing web pages. Following that, the hacker requests money from the victim, typically in Bitcoins. The inability to identify the attacker is the primary factor in choosing the Bitcoin payment mechanism for the ransom. The hacker sends the unlock key if the victim pays the ransom demand. Sometimes, even after receiving money, hackers fail to transmit the decryption key, resulting in permanent data loss. Some modern ransomware attacks replicate themselves to propagate over the network by taking advantage of weaknesses in the operating system [5–9]. The operation of a crypto-ransomware assault is shown in Figure 4.4.

4.3.1 Handling the locker ransomware attack

In a locker ransomware attack, the victim's computer is activated and locked when they click on the attachment or web link. The victim can no longer access the computer without paying the required sum. The victim's device can be used when the required ransom is paid. Even after receiving the requested ransom, the attacker sometimes refuses to unlock the victim's device [5,6]. Figure 4.5 shows the working of the locker ransomware attack.

Figure 4.5 Working method of locker ransomware.

4.4 Research in Ransomware Detection

According to the ransomware detection literature now in print, this study is categorized. The material that is now available is divided into two main groups: analysis and counteraction. Prediction, prevention, and detection were the additional categories used to group countermeasure research for ransomware assaults. In recent years, machine learning and deep learning have significantly contributed to impressive advancements in several study domains. More particularly, ML and DL have contributed to the fantastic boom in artificial intelligence. Search, picture recognition, and decision-making have all benefited from machine learning (ML), a fascinating field of computer science [12].

Furthermore, deep learning is based on robust and extensively applicable models, making it possible to retrieve relevant data for demanding jobs. DL is a helpful addition to cybersecurity analytics, malware detection, classification and analysis, botnet identification and detection, cyber-attack mitigation, intrusion detection and prevention, incident response, network traffic analysis, APT detection, and cybercriminals identification. A taxonomy of the many cybersecurity fields' potential applications for ML models is shown in Figure 4.6.

4.4.1 Ransomware detection methods and techniques

These are "static," "dynamic," and "hybrid" techniques for both detection methods. The information is used to map the area of a cyber-attack and identify the various features and actions of the harmful program. The goal is to stop such cyber-attacks on information systems. Figure 4.8 illustrates the three primary axes of malware detection and analysis. These ransomware detection methods and techniques are shown in Figure 4.7.

4.4.1.1 Static analysis

This method makes use of malware's syntax or structural elements. This method is employed to detect malware before it is run. Malware can be analyzed using static analysis without being launched. The majority of the

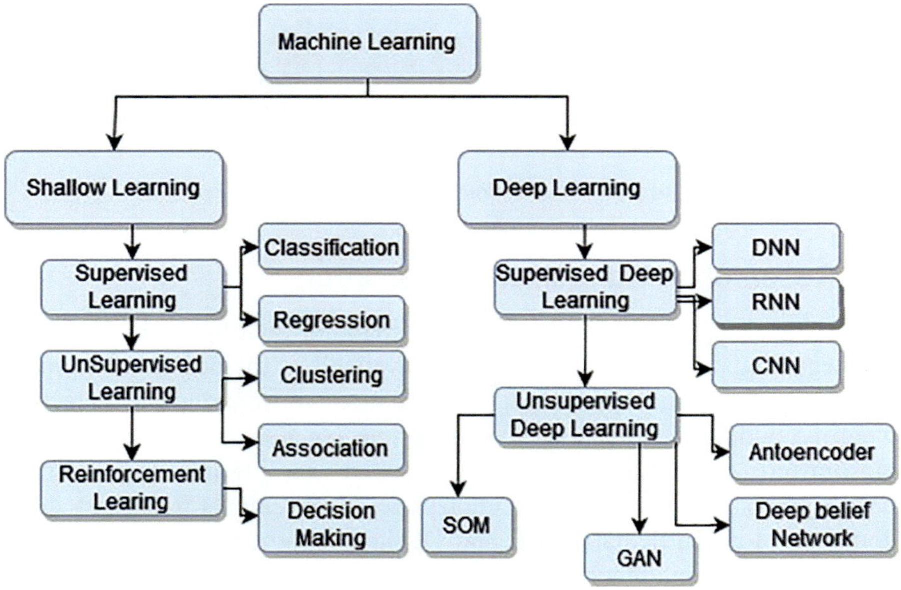

Figure 4.6 Machine-learning-based approaches for cybersecurity.

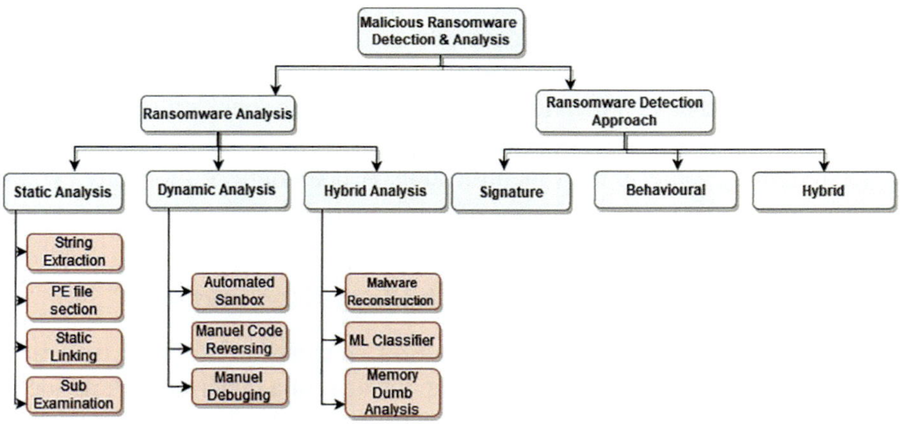

Figure 4.7 Taxonomy of ransomware detection.

metadata must be extracted. While intricate and novel malware cannot be quickly evaluated and discovered using static analysis, old malware can.

4.4.1.2 Dynamic analysis

This method makes use of malware's runtime data. This method is employed to detect malware either during execution or later. Dynamic analysis examines the actions and results of malware while it is being executed. We must comprehend what the malware does once it is run. The objective is to gather real-time information on malware activity and how it affects the system. By employing this technique, we can monitor every part of the malware's activity and how it impacts the neighborhood. The file is frequently executed in a virtual setting. Dynamic analysis is usually preferred over static analysis because malware's characteristics and actions never alter. This makes malware easier to identify, even if its structure changes. Excellent tools for capturing and analyzing packets moving via a network include Wireshark and TCP dump. A sandbox tool for Android apps is called DroidBox.

4.4.1.3 Hybrid analysis

It is the combination of both static and dynamic analyses.

4.4.2 Misuse-based ransomware detection approaches

The signatures of already identified ransomware are employed in misuse-based ransomware detection to infer the existence of potential threats. It uses known and accessible signatures to compare the observed danger with them before making a choice. The two subcategories of misuse detection are structural or signature and behavioral.

4.4.2.1 Signature-based Detection Approaches

The patterns or codes of threats already out there are compared with the code under examination in a signature-based detection approach. The use of out-of-date samples is one of this approach's drawbacks. As time passes, new varieties emerge, necessitating regular updates to the ransomware collection. Overhead was created by this repeated updating [12]. In [14], a signature-based detection method was suggested. This detection strategy used official techniques to find ransomware on mobile devices. The malicious behavior was examined at the bytecode level by this detection mechanism. An android-based ransomware detection system called R-Pack Droid was introduced in [15].

In [16], a strategy for detecting ransomware based on static analysis was suggested. This technique was able to identify malware that can leave environmental fingerprints. In [17], contextual data was used for static analysis.

4.4.3 Behavioral-based detection approaches

Under-examination samples are run in a controlled environment to track their behavior as part of the behavioral-based detection technique. The detection is done according to the behavior that the subject sample(s) exhibit. The study [18] revealed an early warning detection system that monitored ransomware activities on user files.

In [19], a real-time detection technique with early detection was introduced. The research was conducted on the Locky ransomware. The distribution of decoy files and process monitoring were part of the detection system installation. Recursive techniques and a depth-first search were considered for the file traversal.

A ransomware detection technique called self-defensible SSD was put forth in [25] and utilized storage devices for detection. Ransomware was located by transparently monitoring storage access activity, and after it was identified, the garbage collector took care of the backup activities required to preserve the original data. This technique might be used to recover original data without the decryption key. Ransomware was discovered using many read and rewrite/delete requests quickly.

The SSD contained this solution to carry out transparent monitoring. In [26], the ransomware warning was generated via a honeypot approach. Honeypot folders were made, and the changes were watched for. The identification and monitoring were done using Event Sentry and Microsoft File Server Resource Manager. The Microsoft Windows network was under surveillance for a ransomware attack. This technology, however, was unable to stop the start of a ransomware attack. This method was also unreliable because there was no assurance that the assault would always reach the honeypot folders. The administrator deactivated the user account to stop the attack and shut down the machine after it behaved like an alert. Cabaj et al. [20] provided an analysis of the crypto-wall ransomware in a network setting.

4.4.3.1 Hybrid detection approaches

A hybrid approach makes use of both behavioral-based and misuse-based detection techniques. These ransomware detection strategies locate malicious software by combining signature- and behavior-based detection techniques. [21]. A detection method for Android mobile systems called Heldroid [12] employed a misuse-based methodology to carry out the detection. This technique was hybrid since Heldroid ran both a dynamic and statistical analysis on the samples. It was a cutting-edge technique for ransomware samples that still needed to be discovered. The samples were analyzed statically and dynamically as ransomware, goodware, and scareware.

In [22], an automated ransomware detection technique based on deep learning was proposed. This detecting tool was established to help with email filtration. The questionable downloads were used to segregate the emails. Information about ransomware activities was gathered using both static

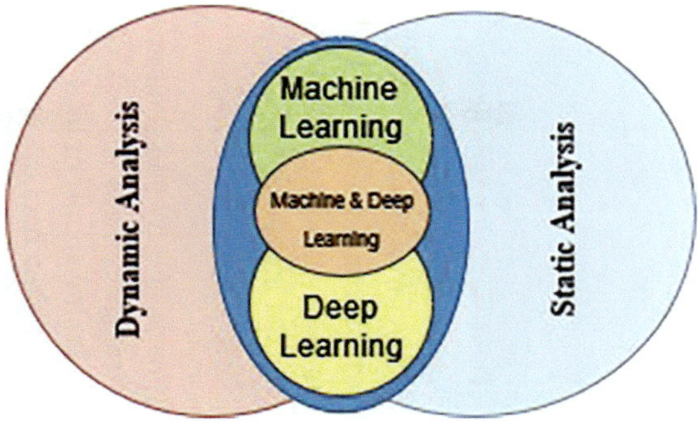

Figure 4.8 Ransomware detection analysis overview.

and dynamic analyses. The proactive monitoring system (PMS) tracked the downloads to find the ransomware attack.

In [23], a PEDA early detection system was suggested to detect ransomware. With the aid of hybrid analysis, two levels were detected. Following a 30-second analysis of the data in the Cuckoo Sandbox, system calls were gathered from the samples. Additionally, this system was able to recognize zero-day attacks. However, the ransomware that employed its encryption code evaded detection by this technology. The feature analysis was completed in [24] using a hybrid analysis. Static and dynamic analyses were conducted on two separate datasets. Ransomware was discovered by extracting static and dynamic features using the feature engineering technique. The ransomware was located using deep convolutional neural networks and machine learning. An overview of different ransomware models is shown in the graphic.

4.5 Surveys on Ransomware DAM (Detection, Avoidance, and Mitigation)

Academics, cybersecurity companies, and governmental organizations have done extensive research into every facet of ransomware dissemination, operation, and creation of effective defense mechanisms. Most frameworks have failed when put into practice, even though organizations and governments have adopted some. This is done for security to encompass a variety of subjects, including network security, data security, application security, and personal cyber-hygiene practices [11][33]. Therefore, developing solutions for comprehensive security takes much work. Numerous research works have looked at the effects of ransomware and gathered information on how to decrease the damage it causes. Our effort is concentrated on offering insights to improve countermeasures in Table 4.1 while summarizing the current detection, avoidance, and mitigation strategies.

Table 4.1 Comprehensive summary of the study.

Authors	Algorithm used	Objective	Pros	Cons
[Amir Djenna et al.] [27]	Deep learning algorithms (CNN and DNN), machine learning algorithms (Random et al.), and heuristic detectors.	To offer a methodical strategy for categorizing contemporary malware utilizing machine and deep learning algorithms on a new comprehensive dataset.	Dynamic deep learning methods and heuristic approaches provide better detection mechanisms for modern malware. Higher accuracy and better performance in detecting modern malware families.	
[Shou-Ching HSIAO et al.] [28]	Reverse engineering, IDA Pro, Windows API (e.g., InternetOpenA, InternetOpenUrlA, LoadResource, etc.).	The authors conducted a static analysis of the infamous WannaCry ransomware, revealing its processes and functions and exploring each chain of malware execution through reverse engineering.	The static analysis allows for a deeper understanding of the inner workings of the WannaCry ransomware, providing valuable insights for future defense strategies and malware detection methods.	The research paper needs to cover the dynamic behavior of the malware, which could also be crucial in understanding its full impact and potential countermeasures.
[Davide Berardi et al.] [29]	Data flooding against ransomware (DFaR) used as a primary technique. The specific tool employed for implementing DFaR is Ranflood, based on the dynamic honeypot approach.	To explore the concept of data flooding against ransomware (DFaR) and introduce a family of techniques, including Ranflood, a dynamic honeypot-based approach to combat ransomware.	DFaR provides a flexible and dynamic approach to combat ransomware. Ranflood and its flooding strategies do not have demanding prerequisites or dependencies, making them easy to implement as drop-in solutions. The dynamic honeypot	For on-the-fly and shadow strategies, there is a risk of losing snapshot lists or shadow backups if targeted by ransomware. Implementing DFaR and Ranflood may require careful consideration of security measures to prevent ransomware from

			approach of DFaR allows monitoring and flooding sensitive locations, increasing detection accuracy.	detecting and encrypting decoy files or snapshots.
[Kenan Begovic et al.] [30]	• I/O monitoring-based detection, file system operations monitoring-based detection, API and system call monitoring-based detection, and machine learning techniques (such as random forest, SVM, KNN, etc.). • I/O monitoring-based detection, file system operations monitoring-based detection, API and system call monitoring-based detection, and machine learning techniques (such as random forest, SVM, KNN, etc.).	The research paper explores various methodologies, including API and system call monitoring, I/O monitoring, and file system monitoring, along with the use of machine learning algorithms, to identify and classify ransomware attacks during the encryption phase, which is a defining characteristic of crypto-ransomware	• API and system call monitoring, I/O monitoring, and file system monitoring are effective techniques for detecting ransomware encryption activities.	Entropy estimation is a promising tool for identifying files encrypted by ransomware, but it may have limitations, such as false positives and vulnerability to certain ransomware evasion techniques.

(*Continued*)

Table 4.1 *Continued*

Authors	Algorithm used	Objective	Pros	Cons
[Bello, Ibrahim, et al.] [31]	Random forest, decision tree, deep learning architecture, Softmax, V-detector negative selection algorithm, gradient tree boosting, complex tree, naive Bayes, multi-layer perceptron.	The paper aims to explore the applications of intelligent algorithms in detecting ransomware attacks and compare their performance in detecting various types of ransomware.	Intelligent algorithms provide a more accurate and efficient way of detecting ransomware attacks. They can classify various types of ransomware and handle large datasets.	Some intelligent algorithms may not be suitable for real-time protection and require complete execution in a controlled environment. Some ransomware may be able to evade detection by the algorithms.
[Simon R. Davies et al.] [32]	• Shannon entropy • Chi-square • Monte Carlo • Serial byte correlation coefficient • Dieharder2 test suite	The research aimed to determine the effectiveness of pure mathematical techniques and statistical test suite analyses in identifying crypto-ransomware encrypted files.	• NIST SP-800 22 tests were universally accepted and used in randomness testing. • Dieharder2 and PractRand also provided helpful results in distinguishing encrypted data. • Shannon entropy and Chi-square were effective in measuring randomness in files.	• The effectiveness of tests may vary depending on the crypto-ransomware strain and encryption method used. • Theoretical entropy reduction techniques used by crypto-ransomware developers may affect detection accuracy.

			• The serial byte correlation coefficient helped detect encryption patterns in files. • Kullback–Liebler divergence was adequate for high entropy files that are not encrypted.	• No crypto-ransomware strain encountered in the study employed entropy reduction techniques.
[Amin Kharraz et al.] [35]	File system activity monitoring Master file table (MFT) analysis I/O request packet (IRP) analysis.	The research aims to analyze and characterize ransomware families based on their malicious activities, explicitly focusing on file system activity, encryption mechanisms, deletion mechanisms, and mitigation strategies. The researchers aim to propose potential defense mechanisms to detect and mitigate ransomware attacks [36].	• The research provides insights into ransomware families' different attack strategies and techniques. • The proposed defense mechanisms, such as file system activity monitoring and decoy resource monitoring, offer potential ways to detect and mitigate ransomware attacks.	The study focuses on a selected set of ransomware families and may not cover all existing variants.

(*Continued*)

Table 4.1 *Continued*

Authors	Algorithm used	Objective	Pros	Cons
[P. Mohan Anand et al.] [37]	Hardware performance counters (HPC)	The research aims to utilize hardware profiling counter registers to detect ransomware, mainly focusing on early detection. The study aims to analyze various ransomware families using HPC and propose a technique for achieving high accuracy in detecting ransomware activity during the initial two seconds of application execution.	• Early detection of ransomware activity can help prevent data encryption and damage. • Utilization of hardware performance counters allows for efficient and effective monitoring of ransomware behavior. • The proposed model achieved high accuracy in detecting ransomware.	• In the future, malware writers can discover ways to avoid HPC-based detection, decreasing the strategy's efficacy. • HPC statistics alone may not be sufficient in cases where ransomware has equivalent statistics to benign applications. • Further research is needed to develop a comprehensive solution that integrates file-based and kernel-based events with hardware profiling counters for improved early ransomware detection.

4.6 Ransomware Prevention Practices

Backup your data: Backing up your data does not prevent ransomware attacks, but it can be the perfect way to recover files. You can wipe the computer clean and restore the backup files with a backup. A recommended approach is the 3-2-1 rule, i.e., consistently backup files and keep three separate copies on two storage types with a copy offline. You can keep any copy on a very secure cloud storage server.

Keep all systems and software updated: For businesses and individuals using onsite physical infrastructure, the chances of preventing ransomware attacks increase by keeping all systems, software, and web browsers up to date. This is important due to the evolving nature of ransomware, as hackers can rely on bugs in non-updated systems to bypass security features and infect a computer. A famous example of ransomware attacks due to outdated systems and software is WannaCry. In 2017, the malicious software crippled the NHS hospitals, Telefonica, and Taiwan Semiconductor Manufacturing Co. (TSMC) by targeting computers with outdated Windows. It affected 230,000 computers for four days.

Install antivirus software and firewalls: Antivirus software and a firewall are two weapons you can use to prevent a ransomware attack. An antivirus works internally (i.e., targeting malware already in the system), while the firewall prevents external attacks. Choosing an antivirus must be a careful process, as only a few antiviruses can detect and respond to malicious software that can constitute a cyber-attack. Also, configuring a firewall can help filter out and block suspicious files from entering the system.

4.6.1 Network segmentation

This is an essential step for businesses whose workers implement a network system (local or remote). Implementing network segmentation in your IT strategy means dividing the organization's network into smaller, isolated, connected networks. This limits the spread of the ransomware in the event of an attack. Implementing a network segmentation strategy for businesses that rely on Salesforce to manage their customer data and workflows can help prevent ransomware attacks and protect critical data.

Each network should have its firewalls and security controls when implementing network segmentation. As a result, this will prevent the ransomware from reaching the target data. Furthermore, it will give your IT team time to identify, isolate, and resolve the attack.

4.6.2 Email protection

Email phishing is one of the significant ways businesses and individuals are exposed to ransomware attacks, with the Federal Bureau of Investigation listing it as a top crime in 2020. Infection can occur via downloading malware masquerading as reputable email attachments, clicking a link to an infected website, and social engineering. You can protect yourself by doing the following [38]:

- Do not open emails, files, and links from unknown sources
- Update all email apps
- Using a sender policy framework (SPF) that authenticates the specific mail servers from which the email is coming
- Using DomainKeys Identified Mail (DKIM), which verifies the ingenuity of the email by providing an encryption key and digital signature
- Using Domain Message Authentication Reporting & Conformance (DMARC)

4.6.3 Application allowlisting

Application and website allowlisting will allow the system to determine the proper application to download and websites to visit on the system or network. Programs and websites not allowed will not be accessible, making downloading infected programs and visiting unsecured ones impossible. Aside from that, some allowlisting software can also block some programs and websites.

4.6.4 Endpoint security

Every device for accessing an organization's files and data is an endpoint with a potential opportunity for criminals to access a network. Consequently, businesses (significantly growing businesses) need to prioritize endpoint security. This is achievable by installing the endpoint detection and response (EDR) or endpoint protection platforms (EPP) for all network users. With both technologies, system administrators can monitor and manage security for each user with a remote device.

4.6.5 Limit user access privileges

Limiting the user access privilege will allow only approved users to access essential data [41]. As a result, you can prevent a ransomware attack from spreading between systems. You can also include a two-factor or multi-factor authentication to prevent access in case there is a security breach.

4.6.6 Conduct routine security testing

New security measures should constantly be put in place. As ransomware strategies continue to adapt to the environment, businesses must regularly conduct cybersecurity testing and assessments. Companies should routinely examine user entitlements and access points, discover new system flaws, and develop new security procedures. Malicious code is tested against current software in a controlled environment as part of the standard approach known as sandbox testing, which assesses the efficacy of security controls [42].

4.6.7 Training on security awareness

Every end user and employer is a point of cyber-attack due to email phishing and social engineering. As a result, you should ensure proper security awareness training to prevent attacks [40]. Basic security training practices you can embark on teaching employees and end users are:

- How to create strong passwords
- How to recognize suspicious emails or attachments
- How to maintain updated systems and software

4.6.8 Switching to cloud computing

Switching to cloud computing involves accessing your business data and files on computer resources provided by a service provider from any internet-enabled device. Aside from cloud computing being performance-focused, providers focus a lot on ensuring high security on their platforms. As a result, the tendency of ransomware attacks is vehemently lower compared to an onsite or local network system. Also, switching to cloud computing is cost-effective as most subscription-based services allow you to focus more on running your organization.

4.6.9 Endpoint security

Endpoint security ought to be a key priority for growing businesses. As organizations expand and end-users increase, more endpoints (laptops, smartphones, servers, etc.) require protection. Thanks to each remote endpoint, criminals may be able to access private data or, worse, the leading network.

4.6.10 Ransomware recovery

You must be prepared to deal with a ransomware assault as soon as it affects your IT resources. A ransomware recovery plan should include the following actions: locate the offending file(s) and find and delete any trigger file(s) from all devices as soon as possible. Choose an assault strategy. The subsequent stages will be determined with the specific kind of ransomware. Screen-locking and encryption-based ransomware are the two main types [39]. Connect none of the gadgets, and disconnect any susceptible devices from your network to stop ransomware from spreading and lessen its consequences.

4.6.11 Ransomware removal

Instructions for uninstalling ransomware that encrypts files. You can follow these instructions to eliminate the encryption ransomware if you were the target of a file encryption ransomware assault [38].

Step 1: Turn off all actual and imagined connections before disconnecting from the internet. Wireless and wired gadgets, external hard drives, numerous storage media, and cloud accounts are among them. Malware cannot now spread over the network as a result of this action. Take the following backup actions in any sites where you suspect there may be additional damage.

Step 2: Use your internet security program to research. Use your installed internet security tool to run a virus scan. You may be aware of the dangers involved. You have the option of quarantining or deleting any harmful files you find. You can manually remove hazardous files using antivirus software or have it do it for you. Manually eliminating the infestation is only advised for expert computer users.

Step 3: Ransomware decryption tool – if ransomware has encrypted your computer's data, the appropriate decryption program will be necessary to regain access. The newest ransomware variants are constantly being researched by Kaspersky so that we can provide the appropriate decryption solutions to protect against these kinds of attacks.

Step 4: Restore your backup – if you have data saved elsewhere or on the cloud, immediately create a backup to prevent ransomware from encrypting. Cleaning and maintaining your computer would be much more difficult without backups. It would help if you regularly created backups to prevent this from happening. Use automated cloud backup services or set up notifications in your calendar if you need to pay more attention to such things.

4.7 Conclusion

A developing area of cybersecurity technology is malware detection and analysis. Malware assaults affect several states and organizations every year. Malware-based cyber-attacks can have catastrophic effects, such as data exfiltration, financial loss, and cyber-espionage. Malware scanners and traditional antiviral measures fall short of protecting users. To do this, practical malware analysis helps create creative strategies to deal with malware events and foresee damage before it occurs. This chapter outlines several operations research techniques for modeling ransomware cybersecurity threats and then discusses the cybersecurity value model framework. A comparative study was made on various proposed ransomware detection techniques and analyzed. As technology develops, this framework's measures, metrics, and best practices may need to be updated and robust. However, the methodology and approach will hold, allowing organizations to adapt and update their security practices to keep up with the times. Exercises for students include researching current measures, metrics, and best practices to become well-versed in current trends. Other exercises include notional scoring for a specific firm or industry. Practitioners can take the framework and directly apply it to their organization, using their firm's data and values to customize and increase security based on its needs. Successful models in practice directly address the problem at hand and are understandable and relatable by decision-makers

References

[1] Kaspersky Security Bulletin (2022). Statistics. Available online: https://securelist.com/ksb-2022-statistics/108129/ (accessed on November 29, 2022).

[2] Seven Biggest Ransomware Attacks in India https://ccoe.dsci.in/resources/ (Published on April 25, 2023).

[3] Malware. Available online: https://www.av-test.org/en/statistics/malware/ (accessed on October 16, 2022).

[4] SonicWall Cyber Threat Report. Available online: https://theblockchaintest.com/uploads/resources/SonicWall%20-%20Cyber%20Threat%20Report%20-%202022%20Feb.pdf (accessed on December 3, 2022).

[5] Harley Malware: New Attack on Android Devices. Available online: https://infosecwriteups.com/harley-malware-new-attackon-android-devices-ae2c599c2217 (accessed on June 5, 2022).
[6] Triada Trojan in WhatsApp Mod. Available online: https://securelist.com/triada-trojan-in-whatsapp mod/103679/ (accessed on June 11, 2022).
[7] Post-Infection Remediation. Available online: https://spycloud.com/ (accessed on October 20, 2022).
[8] The SpyCloud Ransomware Defense Report 2022. Available online: https://spycloud.com/resource/ransomware-defensereport-2022/ (accessed on November 30, 2022).
[9] The Ransomware Threat Landscape. Available online: https://symantec-enterpriseblogs.security.com/blogs/threatintelligence/ransomware-threat-landscape-what-expect-2022 (accessed on October 19, 2022).
[10] Hampton, N., and Baig, Z.A. Ransomware: Emergence of the Cyber-Extortion Menace. In Proceedings of the 13th Australian Information Security Management Conference, Perth, Australia, November 30–December 2, 2015; pp. 47–56. [CrossRef]
[11] Kumar, S., Velliangiri, S., Karthikeyan, P., Kumari, S., Kumar, S., and Khan, M. K. (2021). A survey on the blockchain techniques for the Internet of Vehicles security. Transactions on Emerging Telecommunications Technologies, e4317.
[12] Bernardi, L., Mavridis, T., and Estevez, P. 150 Successful machine learning models: 6 lessons learned at booking.com. In Proceedings of the 25th ACM SIGKDD International Conference on Knowledge Discovery & Data Mining, New York, NY, USA, 4–8 August 2019.
[13] Andronio, N., Zanero, S., and Maggi, F. Heldroid: Dissecting and detecting mobile ransomware. In Proceedings of the International Symposium on Recent Advances in Intrusion Detection, Kyoto, Japan, 2–4 November 2015; pp. 382–404.
[14] Mercaldo, F., Nardone, V., Santone, A., and Visaggio, C.A. Ransomware steals your phone. Formal methods rescue it. In Proceedings of the International Conference on Formal Techniques for Distributed Objects, Components, and Systems, Heraklion, Crete, 6–9 June 2016; Springer: Cham, Switzerland, 2016; pp. 212–221.
[15] Maiorca, D., Mercaldo, F., Giacinto, G., Visaggio, C.A., and Martinelli, F. R-PackDroid: API package-based characterization and detection of mobile ransomware. In Proceedings of the Symposium on Applied Computing, Marrakech, Morocco, 3–7 April 2017; pp. 1718–1723

[16] Zhang, H., Xiao, X., Mercaldo, F., Ni, S., Martinelli, F., and Sangaiah, A.K. Classification of ransomware families with machine learning based on N-Gram of opcodes. Future Gener. Comput. Syst. 2019, 90, 211–221.

[17] Zhang, B., Xiao, W., Xiao, X., Sangaiah, A.K., Zhang, W., and Zhang, J. Ransomware classification using patch-based CNN and self-attention network on embedded N-grams of opcodes. Future Gener. Comput. Syst. 2020, 110, 708–720.

[18] Scaife, N., Carter, H., Traynor, P., and Butler, K.R. Cryptolock (and drop it): Stopping ransomware attacks on user data. In Proceedings of the 2016 IEEE 36th International Conference on Distributed Computing Systems (ICDCS), Nara, Japan, 27–30 June 2016; pp. 303–312.

[19] Feng, Y., Liu, C., and Liu, B. Poster: A new approach to detecting ransomware with deception. In Proceedings of the 38th IEEE Symposium on Security and Privacy Workshops, San Jose, CA, USA, 22–24 May 2017.

[20] 76. Cabaj, K., Gawkowski, P., Grochowski, K., and Osojca, D. Network activity analysis of CryptoWall ransomware. Prz. Elektrotech. 2015, 201–204.

[21] Damodaran, A., Di Troia, F., Visaggio, C.A., Austin, T.H., and Stamp, M. A comparison of static, dynamic, and hybrid analysis for malware detection. J. Comput. Virol. Hacking Tech. 2017, 13, 1–12.

[22] Lokuketagoda, B., Weerakoon, M.P., Kuruppu, U.M., Senarathne, A.N., and Abeywardena, K.Y. R-Killer: An email-based ransomware protection tool. In Proceedings of the 2018 13th International Conference on Computer Science & Education (ICCSE), Colombo, Sri Lanka, 8–11 August 2018; pp. 1–7.

[23] Kok, S., Abdullah, A., and Jhanjhi, N. Early detection of crypto-ransomware using pre-encryption detection algorithm. J. King Saud Univ.-Comput. Inf. Sci. 2020, in press.

[24] Ashraf, A., Aziz, A., Zahoora, U., Rajarajan, M., and Khan, A. Ransomware Analysis using Feature Engineering and Deep Neural Networks. arXiv 2019, arXiv:1910.00286.

[25] Paik, J.Y., Shin, K., and Cho, E.S. Poster: Self-defensible storage devices based on flash memory against ransomware. In Proceedings of the IEEE Symposium on Security and Privacy, San Jose, CA, USA, 23–25 May 2016.

[26] Moore, C. Detecting ransomware with honeypot techniques. In Proceedings of the 2016 Cybersecurity and Cyberforensics Conference (CCC), Amman, Jordan, 2–4 August 2016; pp. 77–81.

[27] Djenna, A., Bouridane, A., Rubab, S., and Marou, I.M. Artificial Intelligence-Based Malware Detection, Analysis, and Mitigation. Symmetry 2023, 15, 677. https://doi.org/10.3390/sym15030677

[28] S.C. Hsiao and D.Y. Kao. "The static analysis of WannaCry ransomware," 2018 20th International Conference on Advanced Communication Technology (ICACT), Chuncheon, Korea (South), 2018, pp. 153–158, doi: 10.23919/ICACT.2018.8323680.

[29] Berardi, Davide, et al. "Data Flooding against Ransomware: Concepts and Implementations." Computers & Security 131 (2023): 103295.

[30] Kenan Begovic, Abdulaziz Al-Ali, and Qutaibah Malluhi. "Cryptographic ransomware encryption detection: Survey", Computers & Security, Volume 132, 2023, 103349, ISSN 0167-4048,https://doi.org/10.1016/j.cose.2023.103349.

[31] Bello, Ibrahim, et al. "Detecting ransomware attacks using intelligent algorithms: Recent development and next direction from deep learning and big data perspectives." Journal of Ambient Intelligence and Humanized Computing 12 (2021): 8699–8717.

[32] Davies, S.R., Macfarlane, R., and Buchanan, W.J. Comparison of Entropy Calculation Methods for Ransomware Encrypted File Identification. Entropy 2022, 24, 1503. https://doi.org/10.3390/e24101503

[33] Ba'abbad, I., and Batarfi, O. Proactive Ransomware Detection Using Extremely Fast Decision Tree (EFDT) Algorithm: A Case Study. Computers 2023, 12, 121. https://doi.org/10.3390/computers12060121

[34] Umar, R., Riadi, I., and Kusuma, R.S. (2021). Mitigating Sodinokibi Ransomware attack on cloud network using Software-Defined Networking (SDN). International Journal of Safety and Security Engineering, Vol. 11, No. 3, pp. 239–246. https://doi.org/10.18280/ijsse.110304

[35] Pauch, Dariusz. "Ransomware Attacks As A Cybersecurity Insurance Coverage Threat." Humanities and Social Sciences 30.2 (2023): 99–107.

[36] Bold, R., Al-Khateeb, H., and Ersotelos, N. Reducing False Negatives in Ransomware Detection: A Critical Evaluation of Machine Learning Algorithms. Appl. Sci. 2022, 12, 12941. https://doi.org/10.3390/app122412941

[37] Mohan Anand, P., Sai Charan, P.V., and Sandeep, K Shukla. (2023). HiPeR - Early Detection of a Ransomware Attack using Hardware Performance Counters. Digital Threats Just Accepted (July 2023). https://doi.org/10.1145/3608484

[38] Han, W., Xue, J., Wang, Y., Zhang, F., and Gao, X. APTMalInsight: Identify and cognize APT malware based on system call information and ontology knowledge framework. Inf. Sci. 2021, 546, 633–664.

[39] Almashhadani, A.O., Carlin, D., Kaiiali, M., and Sezer, S. MFMCNS: A multi-feature and multi-classifier network-based system for ransomware detection. Comput. Secure. 2022, 121, 102860.

[40] Beaman, C., Barkworth, A., Akande, T.D., Hakak, S., and Khan, M.K. Ransomware: Recent advances, analysis, challenges, and future research directions. Comput. Secure. 2021, 111, 102490.

[41] Khammas, B.M. Ransomware detection using random forest technique. ICT Express 2020, 6, 325–331.

[42] Fernando, D.W., and Komninos, N. FeSA: Feature selection architecture for ransomware detection under concept drift. Comput. Secure. 2022, 116, 102659.

5

Cyber Attacker Profiling and Cyberbullying Overview

Wasswa Shafik

Dig Connectivity Research Laboratory (DCRLab), Kampala, Uganda

Email: wasswashafik@ieee.org

Abstract

Because cyberbullying is more common as different devices connect to the Internet, people are becoming more interested in finding cyber attackers and planning to stop and lessen these attacks. Cyber attacker profiling has also provided an exciting way to learn about cyber attackers' targets, goals, techniques, and behavior patterns. By analyzing the characteristics and patterns of behavior of cyber attackers, it is possible to identify potential attackers before they carry out an attack and to develop effective prevention and response strategies, for example, employee training, implementing strict access controls to limit the information available to employees based on their roles, intrusion prevention systems (IPS), intrusion detection systems (IDS), regular audits and penetration testing, and privacy settings. However, there are concerns about the ethical implications of profiling and the potential for bias and discrimination, including privacy and security balance, transparency and explainability, false positives and due process, and regulatory and legal compliance. This study depicts an overview and explores the current research on cyber attacker profiling and cyberbullying, including the challenges and opportunities associated with these approaches. In addition, the study highlights the need for further research and collaboration to develop more effective and ethical strategies for preventing and responding to cyberbullying and other types of cyber-attacks. Finally, this research provides insight into a possible path that can be taken to solve a number of the concerns. Cyber-related crime remains prevalent in the communication sphere, posing a significant threat to all stakeholders' online security and privacy.

5.1 Introduction

Cyberbullying has become increasingly prevalent with the rise of social media and other online platforms. Cyberbullying uses technology to harass, intimidate, or harm individuals or groups. It can take many forms, including sending threatening messages, spreading rumors or lies, sharing private photos or videos, and creating fake profiles or accounts. Cyber-attacks are another growing concern in the digital age [1]. These attacks involve using technology to gain unauthorized access to systems, networks, or devices, steal data, disrupt operations, or cause damage. Cyber attackers can be individuals, groups, or even state-sponsored entities, and they use various techniques to carry out their attacks, including phishing, malware, and social engineering [2].

Both cyberbullying and cyber-attacks have significant impacts on individuals, organizations, and society as a whole. For example, cyberbullying can lead to psychological harm, social isolation, and even suicide, while cyber-attacks can cause financial losses, reputational damage, and even physical harm in some cases. As a result, there has been a growing interest in developing strategies to prevent and mitigate these risks. One promising approach to understanding cyber attackers and preventing cyber-attacks is cyber-attack profiling. Cyber attacker profiling involves analyzing the characteristics, motivations, and patterns of behavior of cyber attackers to identify potential attackers before they carry out an attack [3]. This approach has the potential to provide valuable insights into the mindset and tactics of cyber attackers and to develop effective response strategies.

However, there are concerns about the ethical implications of cyber-attacked profiling, including the potential for bias and discrimination. This study will explore the current state of research on cyber attacker profiling and cyberbullying, including the challenges and opportunities associated with these approaches. We will also discuss the need for further research and collaboration to develop more effective and ethical strategies for preventing and responding to cyberbullying and other cyber-attacks.

Cyber attacker profiling involves analyzing the characteristics and patterns of the behavior of cyber attackers to identify potential attackers before they carry out an attack. This approach can help security professionals and law enforcement agencies understand cyber attackers' motivations, techniques, and tactics and develop effective response and prevention approaches. One of the key benefits of cyber attacker profiling is that it can help identify patterns of behavior that may indicate an impending attack [4]. Suppose a cyber attacker typically targets a specific organization or system or uses a particular technique or tool. In that case, this information can be used to anticipate and prevent future attacks.

Cyber attacker profiling can also help identify the motivations and goals of cyber attackers. For example, some attackers may be motivated by financial gain, while others may seek to disrupt or undermine a particular organization or government. Understanding these motivations can help to inform prevention and response strategies and to develop more effective measures to deter future attacks. However, there are also challenges associated with cyber attacker profiling. One of the main challenges is the ethical implications of profiling, particularly in terms of bias and discrimination [5]. Profiling can lead to unfair treatment of individuals or groups based on race, ethnicity, gender, or other characteristics. It is, therefore, essential to ensure that profiling is conducted ethically and unbiasedly and that appropriate safeguards are in place to prevent discrimination.

Another challenge is the difficulty of identifying and tracking cyber attackers. Cyber attackers can be difficult to identify, mainly by using techniques such as spoofing or masking their internet protocol (IP) address. Furthermore, cyber attackers may use multiple devices or locations to carry out attacks, making it difficult to track their movements [6]. Despite these challenges, cyber attacker profiling can provide valuable insights into cyber attackers' motivations and tactics and develop operative preventions and response stratagems.

While cyber-attacks target organizations or systems, cyberbullying is directed toward individuals or groups and can severely affect mental and emotional health, social relationships, and even physical safety. Cyberbullying can occur through various digital platforms, including social media, email, text messages, and online gaming. One of the challenges in addressing cyberbullying is the difficulty of identifying and tracking perpetrators [7]. Cyberbullies may use fake profiles or anonymous accounts to avoid detection, making it challenging for law enforcement and school officials to identify and act against them. Additionally, cyberbullying often occurs outside of traditional school or workplace environments, making it difficult for teachers, employers, or other authority figures to intervene.

Moreover, cyberbullying can also have a significant impact on mental and emotional well-being. Studies have shown that victims of cyberbullying are at increased risk for depression, anxiety, and suicidal ideation. This underscores the need for effective prevention and intervention strategies to address the issue. While cyber attacker profiling and cyberbullying are distinct issues, there are opportunities for collaboration between the two fields [8]. Understanding the characteristics and patterns of behavior of cyberbullies can help identify potential cyber attackers who may be motivated by similar factors, such as a desire for power or control.

Effective prevention and intervention strategies for cyberbullying and cyber-attacks should also consider the ethical implications of profiling. It is crucial to ensure that any profiling is conducted ethically and unbiasedly and that appropriate safeguards are in place to prevent discrimination [8]. In addition to profiling, other strategies can be employed to prevent and mitigate cyberbullying and cyber-attacks. Increasing public awareness about the risks and consequences of cyberbullying and cyber-attacks can help promote safe and responsible online behavior.

The importance of addressing cyberbullying and cyber-attacks is paramount in the context of our quickly advancing technology environment. The growing reliance of society on cyberspace for communication, trade, and social contact underscores the imperative to safeguard persons and organizations from these vulnerabilities. Cyber-attacks have the potential to cause significant financial damage, compromise data security, and disrupt essential infrastructure. Similarly, cyberbullying can have severe psychological and emotional consequences, especially among the younger population, such as children and adolescents [8].

The widespread adoption of digital devices and online platforms has accumulated and retained substantial personal data. It is crucial to combat cyber-attacks to safeguard individuals' privacy and the confidentiality of their sensitive information [8]. Furthermore, it is critical to note that cyber-attacks significantly impact national security, as they can specifically target critical government systems and defense infrastructure. Implementing measures to prevent such attacks is crucial to safeguard a nation's interests and maintain its sovereignty.

The imperative of tackling cyberbullying lies in its significance for upholding social cohesiveness and cultivating a digital milieu that promotes respect and inclusivity. Cyberbullying can have significant psychological ramifications, disrupting social harmony, creating splits within groups, and enduring consequences for individuals' emotional well-being [9]. Establishing and preserving confidence in developing technologies hinges upon the assurance of security and safety within the digital domain [10]. It becomes imperative to tackle the issues of cyberbullying and cyber-attacks in order to foster and sustain trust, as these factors significantly influence technological trust and adoption.

The remainder of this study is structured as follows. In Section 5.2, the background of cyberbullying and cyber attacker profiling entailing definition of cyberbullying and types of cyberbullying like verbal, social, physical, psychological, sexual, and stalking cyberbullying, and anti-cyberbullying registration are discussed. Section 5.3 illustrates different forms of cybercrime, including phishing, malware, identity theft, ransomware, cyberbullying, and online scams. In Section 5.4, attacker profiling is presented and explained, including high- and low-level attacker attributes. Finally, in Section 5.5, attacker profiling data sources are demonstrated, including attack history, threat intelligence, vulnerability databases, network traffic data, malware analysis, social media monitoring, incident response data, insider threat data, and physical security data, among others. Security measures are presented in Section 5.6 and the conclusion in Section 5.7.

5.2 Background

5.2.1 Cyberbullying definition

Cyberbullying, sometimes called cyber harassment, refers to using digital technologies, such as social media, messaging apps, or online forums, to harass, intimidate, or harm individuals or groups. Cyberbullying can take many forms, including sending threatening or insulting messages, spreading rumors or false information, or sharing embarrassing photos or videos without consent [11]. It can severely affect mental and emotional health, social relationships, and even physical health. It can be particularly challenging due to the anonymity and widespread reach of digital communication.

5.2.2 Types of cyberbullying

Cyberbullying can take many different forms and can be broadly classified into the following types.

5.2.2.1 Verbal cyberbullying

This type of cyberbullying uses language to harm or intimidate the victim. This can include sending threatening or insulting messages, spreading rumors or false information, or using derogatory language to demean the victim.

5.2.2.2 Social cyberbullying

This involves using social media platforms, such as Facebook, Twitter, or Instagram, to spread rumors or negative comments about the victim or to isolate them from their peers.

5.2.2.3 Physical cyberbullying

This type of cyberbullying involves using digital technology to harm the victim physically. This can include posting photos or videos of the victim being physically attacked or hurt or encouraging others to engage in physical harm.

5.2.2.4 Psychological cyberbullying

This involves using digital technology to harm the victim's mental and emotional well-being. This can include posting derogatory or hurtful comments or spreading rumors or false information to damage the victim's reputation.

5.2.2.5 Sexual cyberbullying

This involves using digital technology to engage in sexually explicit behavior, such as sharing intimate photos or videos without consent or using sexually explicit language to harass or intimidate the victim.

5.2.2.6 Stalking cyberbullying

This involves using digital technology to track and monitor the victim's online activities or to threaten or harass them in person. It is important to note that these types of cyberbullying are not mutually exclusive and often overlap [12], [13]. Victims of cyberbullying may experience multiple types of bullying at the same time, which can have severe consequences for their mental and emotional well-being. Cyberbullying's adverse effects are illustrated in Figure 5.1.

5.2.3 Anti-cyberbullying regulation

Anti-cyberbullying regulation is a policy approach that seeks to prevent or reduce the incidence of cyberbullying and other online harms by requiring users to register their personal information when using online services or platforms. This approach is based on the idea that requiring users to provide personal information, such as their name, age, and contact information, will make them more accountable for their actions online and deter them from engaging in harmful behaviors [14]. Proponents of anti-cyber registration argue that it can help address the problems of anonymity and accountability online. Requiring users to provide their real identities can make it easier for law enforcement agencies to track and investigate online crimes, including cyberbullying and harassment. Additionally, it can help to promote responsible online behavior, as users will be more aware of the potential consequences of their actions and less likely to engage in harmful behavior [15].

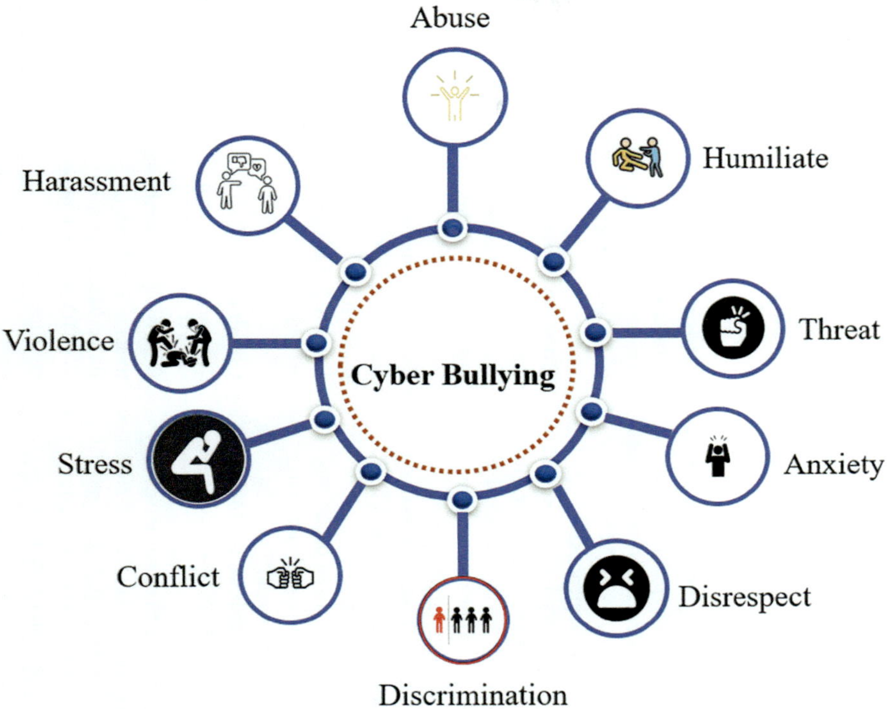

Figure 5.1 Cyberbullying's adverse effects.

However, critics of anti-cyber registration argue that it can have negative consequences for privacy and freedom of expression. Requiring users to provide personal information may discourage some individuals from using online services or platforms, particularly those concerned about the potential misuse or theft of their data [16]. Additionally, it may stifle free speech and expression online, as individuals may hesitate to express their opinions or share controversial content if required to provide their real identity. While anti-cyber registration may have some potential benefits in promoting responsible online behavior and addressing online harms, it is vital to consider such policies' potential drawbacks and unintended consequences [17]. Table 5.1 presents a comparative summary that elucidates significant distinctions between cyber assaults and cyberbullying about their purpose, victims, methodologies, incentives, circumstances, and legal and ethical ramifications.

In order to effectively combat cyberbullying and other forms of online harm, it is crucial to adopt an approach that upholds the principles of privacy and freedom of expression while also relying on evidence-based measures

Table 5.1 A comparative summary of cyber-attacks and cyberbullying.

Trait	Cyber-attacks	Cyberbullying	Impact	Case	Reference
Intent	Malicious intent to compromise systems, data, or resources	Intent to harass, intimidate, or harm individuals online	Financial loss, data breaches, disruption, psychological distress	Ransomware attack on a hospital	[29]
Target	Individuals, organizations, systems, data	Primarily individuals, often in social or school settings	Privacy invasion, emotional harm, reputational damage	Persistent online harassment of a teenager	[30, 31]
Techniques	Phishing, malware, DDoS attacks, ransomware	Harassment, trolling, doxing, impersonation, exclusion, etc.	System compromise, data theft, social isolation, social consequences	Impersonating someone on social media	[32]
Motivation	Financial gain, espionage, political, or ideological	Personal grudges, personal satisfaction, peer pressure	Financial losses, intellectual property theft, fear, anxiety	Extortion through online threats	[33]
Context	Often broader and may involve nation-states or crime rings	Typically occurs within online communities or social networks	Can escalate to legal action, disciplinary measures, or criminal charges	Cyber-attack on a critical infrastructure	[34]
Legal and ethical implications	May lead to legal consequences, regulatory fines	Legal consequences, school interventions, online platform policies	Legal action, loss of online privileges, mental health support	Cyberbullying case leading to school intervention	[35]
Frequency	It can be sporadic or targeted, depending on the objectives	It can be persistent, leading to ongoing harm and distress	Varies from occasional incidents to continuous harassment	Repeated email phishing campaigns	[36]
Complexity	Often requires technical expertise and resources	Primarily involves online interactions and social dynamics	Depends on the sophistication of the tactics used	Coordinated cyber-espionage campaign	[37]
Reporting	It may not be immediately detected or reported	Often reported by victims, parents, or concerned peers	Victims or witnesses may report incidents to authorities	Reporting cyberbullying to school authorities	[38]
Response time	Rapid response is critical to mitigate damage	Timely intervention and support can reduce harm	Prompt reporting and intervention can limit emotional distress	Immediate response to a DDoS attack on a website	[39]

that have proven to be effective in mitigating online harm. Table 5.1 highlights the diverse nature of cyber threats and their unique consequences for both persons and organizations.

5.3 Forms of Cybercrime

In this section, we present some particular forms of cybercrime.

5.3.1 Phishing

Phishing is tricking someone into sharing sensitive information, such as passwords or credit card numbers, by posing as a trustworthy entity in an email or message [18]. The solution to this problem is to educate people about the dangers of phishing and how to identify fake emails. In addition, future research could focus on developing more advanced algorithms to detect and prevent phishing attacks.

5.3.2 Malware

Malware is software designed to harm or disrupt computer systems, often by stealing data or installing viruses [19]. The solution to this problem is to use antivirus software and keep it up-to-date. Future research could focus on developing more sophisticated methods of detecting and preventing malware.

5.3.3 Identity theft

Identity theft is using someone's personal information to commit fraud or other crimes. The solution to this problem is to protect personal information, such as passwords and social security numbers, and monitor credit reports for any suspicious activity [20]. Future research could focus on developing more secure authentication methods, such as biometric identification.

5.3.4 Ransomware

Ransomware is malware that encrypts files on a computer system, making them inaccessible to the user and demanding payment in exchange for a decryption key [21]. The solution to this problem is to back up important data regularly and to use antivirus software. Future research could focus on developing new ways to prevent and recover from ransomware attacks.

5.3.5 Cyberbullying

Cyberbullying is using technology to harass, intimidate, or embarrass someone. The solution to this problem is to educate people about the dangers of cyberbullying and to encourage online civility. In addition, future research could focus on developing new strategies for preventing and responding to cyberbullying.

5.3.6 Online scams

Online scams are schemes designed to defraud people by promising something too good to be true, such as a lottery win or a job offer. The solution to this problem is to be cautious and skeptical of unsolicited emails or messages and verify the authenticity of any offers before providing personal information or payment [22]. Future research could focus on developing more advanced methods of detecting and preventing online scams.

5.3.7 Hacking

Hacking is unauthorized access to a computer system or network, often intending to steal data or cause damage. The solution to this problem is using strong passwords, keeping software up-to-date, and firewalls and other security measures [23]. Future research could focus on developing more advanced methods of detecting and preventing hacking attacks.

5.3.8 Cyberstalking

Cyberstalking is using technology to harass or intimidate someone, often by sending threatening or abusive messages. The solution to this problem is to report any instances of cyberstalking to law enforcement and block or ignore the perpetrator. In addition, future research could focus on developing new ways to prevent and respond to cyberstalking.

5.3.9 Denial-of-service (DoS) attacks

DoS attacks are designed to overwhelm a computer system or network with traffic, making it unavailable to users [24]. The solution to this problem is to use firewalls and other security measures to prevent such attacks. In addition, future research could focus on developing new ways to detect and respond to DoS attacks.

5.3.10 Cyber-espionage

Cyber-espionage uses technology to steal sensitive information from a company or organization, often to gain a competitive advantage. The solution to this problem is to use encryption and other security measures to protect sensitive data and monitor networks for suspicious activity [24]. Future research could focus on developing more advanced methods of detecting and preventing cyber-espionage.

5.4 Attacker Profiling

Attacker profiling is the process of analyzing the characteristics and behavior of cyber attackers to identify patterns and develop strategies for defending against them. This approach involves collecting and analyzing attackers' data, motivations, tactics, and techniques to gain insights into their behavior and identify vulnerabilities in systems that attackers may exploit [25]. The goal of attacker profiling is to create a comprehensive picture of an attacker's behavior and motivations, anticipate their actions, and develop effective defense strategies. This process typically involves collecting data from various sources, including network logs, threat intelligence reports, and other sources of information about cyber-attacks.

Once data has been collected, it can be analyzed to identify patterns in attackers' behavior, such as the tools and techniques they use, the types of systems they target, and the methods they use to evade detection. This information can then be used to develop defense strategies, such as configuring firewalls and intrusion detection systems to block known attack patterns and vulnerabilities. Attacker profiling is an ongoing process that requires continuous data collection and analysis to stay up-to-date with the latest threats and attack techniques [26]. Maintaining a multidisciplinary approach involving experts in network security, data analytics, and behavioral psychology is also important.

5.4.1 High-level attacker's attribute

Attacker profiling involves the analysis of a variety of high-level attributes to gain insights into the behavior and motivations of cyber attackers [27], as Figure 5.2 demonstrates. Some of the critical high-level attributes that are commonly considered in attacker profiling include the following.

5.4.1.1 Motivations

Understanding an attacker's motivations is a crucial aspect of attacker profiling. This may include financial gain, political or ideological motives, or simply the thrill of the attack.

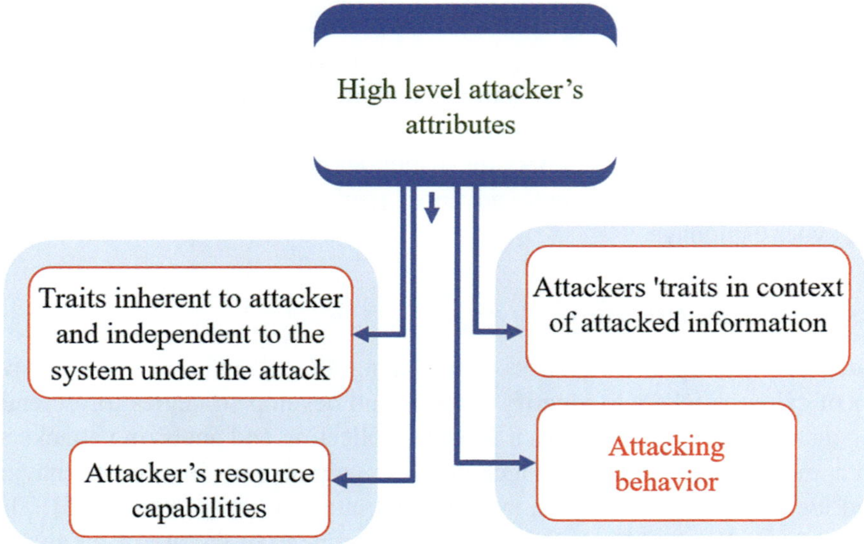

Figure 5.2 High-level attribute groups.

5.4.1.2 Tools and techniques

Attackers typically use various tools and techniques to gain access to systems and networks. Therefore, understanding the tools and techniques commonly used by attackers can help organizations better defend against them.

5.4.1.3 Tactics and procedures

Attackers often follow specific procedures or tactics to achieve their goals. Understanding these tactics and procedures can help organizations anticipate an attacker's next move and develop effective defense strategies.

5.4.1.4 Target selection

Attackers may target specific systems or networks depending on their motivations and goals. Understanding an attacker's target selection criteria can help organizations protect their most valuable assets better.

5.4.1.5 Technical expertise

Attackers may have varying technical expertise, ranging from highly skilled professionals to less sophisticated attackers who use off-the-shelf tools and scripts. Understanding an attacker's technical capabilities can help organizations develop more effective defense strategies.

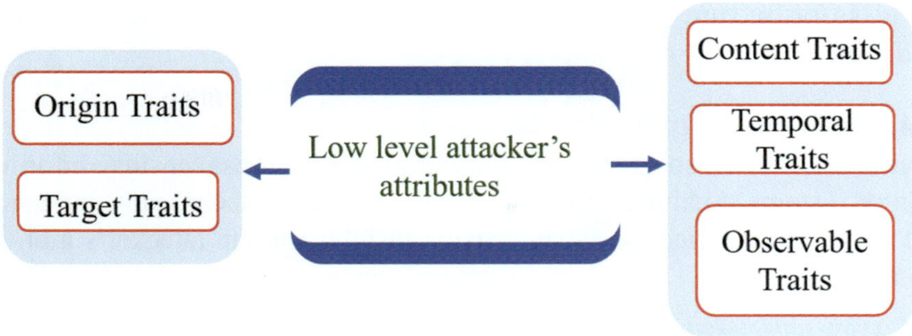

Figure 5.3 Low-level attribute groups.

5.4.1.6 Attribution

Attacker profiling may also involve identifying the individual or group behind a specific attack. This may involve analysis of the attack and other sources of information such as social media and online forums. Inclusively, these high-level attributes can provide valuable insights into the behavior and motivations of attackers, allowing organizations to better anticipate and defend against cyber threats, as Figure 5.3 demonstrates.

5.4.2 Low-level attribute groups

In addition to high-level attributes, attacker profiling also involves the analysis of low-level attributes to gain a more detailed understanding of an attacker's behavior and techniques [28]. Some of the critical low-level attributes that are commonly considered in attacker profiling include the following.

5.4.2.1 Network traffic

Analyzing network traffic can provide valuable insights into an attacker's activity, including the tools and techniques they are using, the systems they are targeting, and the data they are exfiltrating.

5.4.2.2 Malware

Malware is a standard tool attackers use to access systems and networks. Analyzing the characteristics of malware, including its code structure and behavior, can provide insights into the identity and behavior of the attacker.

5.4.2.3 Command and control infrastructure

Attackers often use a command-and-control infrastructure to communicate with malware and control its behavior. Therefore, analyzing the characteristics of this infrastructure can provide valuable insights into an attacker's behavior and motivations.

5.4.2.4 Social engineering techniques

Many attackers use social engineering techniques to access sensitive information or systems. Analyzing the characteristics of these techniques, including the language and tone used, can provide insights into an attacker's motivations and goals.

5.4.2.5 Exploit code

Attackers often exploit code to exploit vulnerabilities in systems and networks. Analyzing the characteristics of exploit code, including its code structure and behaviors, can provide insights into an attacker's technical capabilities and motivations.

5.4.2.6 Indicators of compromise

Attacker profiling may also involve the analysis of indicators of compromise, which are specific artifacts or patterns of behavior that indicate an attacker's presence. Analyzing these indicators can provide valuable insights into an attacker's behaviors and tactics. These low-level attributes can provide detailed insights into an attacker's behaviors and techniques, allowing organizations to better understand and defend against cyber threats [29]. However, analyzing these attributes requires specialized technical expertise and advanced tools and may require significant resources and investment.

The links between high-level and low-level attributes in attacker profiling can be complex, as multiple low-level attributes may be associated with a single high-level attribute. Conversely, a single low-level attribute may be associated with multiple high-level attributes, as shown in Figure 5.4. Nevertheless, understanding these links is vital to developing effective attacker profiling strategies.

For example, suppose an attacker's high-level attribute is identified as being motivated by financial gain. In that case, the related low-level attributes may include ransomware or other malware designed to steal or extort money [30]. Similarly, suppose an attacker's high-level attribute is identified as being technically sophisticated. The related low-level attributes may include advanced exploit code, custom malware, or a complex command and control infrastructure.

Conversely, analyzing low-level attributes may provide insights into an attacker's high-level attributes. For example, suppose an attacker is found to be using social engineering techniques to gain access to sensitive information or systems. In that case, this may indicate a high-level attribute of being

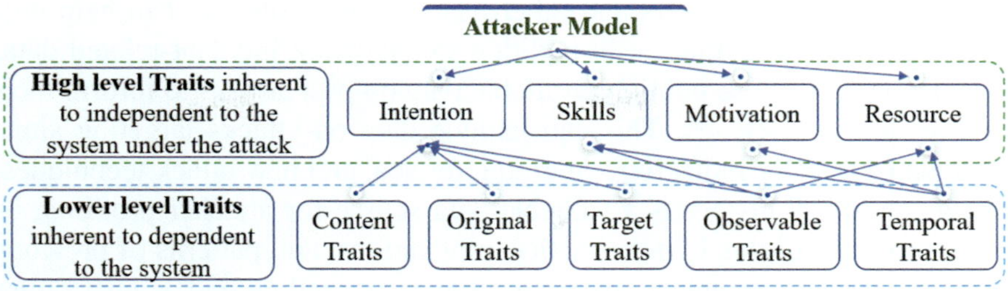

Figure 5.4 High- and low-level attributes link synopsis.

motivated by the desire to steal intellectual property or gain a competitive advantage. Similarly, suppose an attacker is found to be using a specific type of malware that is commonly associated with a particular threat actor group. In that case, this may indicate a high-level attribute of being motivated by political or ideological goals [31]. The links between high-level and low-level attributes in attacker profiling are complex and dynamic and require a nuanced and multidisciplinary approach to be fully understood and utilized for effective defense against cyber threats.

5.5 Attacker Profiling Data Sources

Attacker profiling analyzes data to create a profile of a potential attacker, including their characteristics, motivations, and potential attack methods. The following are 10 familiar data sources used in attacker profiling.

5.5.1 Attack history

Attack history refers to analyzing data related to previous attacks to identify patterns, trends, and commonalities among attackers and their methods. Attack history data sources may include information from past incidents that an organization has experienced and public information on attacks targeting similar organizations or industries [32]. By analyzing attack history, an organization can develop an understanding of the tactics, techniques, and procedures (TTPs) used by attackers, including their methods of gaining access to systems, the types of data they target, and their motivations. This information can be used to develop threat models, identify potential attack vectors, and inform security controls and incident response plans.

For example, an organization may analyze previous phishing attacks and identify commonalities among the emails, such as similar subject lines or sender addresses. This information can be used to inform employee training and to develop email filters and other technical controls that can help prevent future phishing attacks. In addition to analyzing incident-related data, organizations may use publicly available data on past attacks to inform their attacker profiling efforts. This can include data on attacks targeting similar organizations or industries, emerging threats, and new attack techniques. Analyzing attack history is a valuable data source for attacker profiling, as it can provide insights into the tactics, motivations, and patterns of previous attackers, helping organizations better prepare for and defend against future attacks. In addition, analyzing the historical data of previous attacks can provide insights into the tactics and methods used by attackers.

5.5.2 Threat intelligence

Threat intelligence is the process of gathering, analyzing, and sharing information on current and emerging cyber threats. This data source is a crucial component of attacker profiling, as it helps organizations understand the current threat landscape and anticipate potential attacks. It may include information on known threat actors, their motivations, tactics, targets, and data on emerging threats and vulnerabilities [32]. This data can be gathered from various sources, including open-source intelligence, dark web monitoring, security research reports, and information-sharing platforms.

By analyzing threat intelligence data, organizations can develop an understanding of the tactics and methods used by attackers, including the tools and techniques they use to gain access to systems and data. This information can be used to inform threat models, identify potential vulnerabilities, and develop security controls that can help prevent or mitigate attacks. For example, if a threat intelligence report identifies a new strain of malware targeting a specific industry, an organization can use this information to inform its attacker profiling efforts [33]. They can better understand the malware's characteristics, such as how it spreads and its potential impact on their systems, and take steps to prevent infection, such as implementing additional security controls or updating their antivirus software. Information about current and emerging threats can help anticipate potential attacks and inform risk management strategies.

5.5.3 Vulnerability databases

Vulnerability databases are data sources used in attacker profiling to identify potential weaknesses in a system or network. These databases contain known vulnerabilities, including details about how they can be exploited and the potential impact of an exploit. Vulnerability databases are typically maintained by security researchers, vendors, and other organizations that specialize in identifying and mitigating security vulnerabilities. Some common vulnerability databases include the national vulnerability database, the common vulnerabilities and exposures database, and the open web application security project top 10 list [33].

By analyzing vulnerability databases, organizations can develop an understanding of potential attack vectors and prioritize their efforts to address vulnerabilities that pose the most significant risk. For example, if a vulnerability database identifies a critical vulnerability in a widely used software application, an organization can prioritize patching or updating that application to prevent exploitation. In addition to analyzing vulnerability databases for known vulnerabilities, organizations can use this data source to identify emerging threats and new attack techniques. Organizations can better anticipate potential attacks and develop adequate security controls to mitigate risk by staying current on the latest vulnerabilities and exploits. Analyzing publicly available vulnerability databases can help identify weaknesses that attackers may exploit.

5.5.4 Network traffic data

Network traffic data is a data source used in attacker profiling that involves analyzing the traffic that flows across a network to identify potential threats and anomalous behavior. Network traffic data can include information such as the source and destination IP addresses of network packets, the type of traffic (like web traffic, email traffic, and file transfers), and the size and content of network packets. By analyzing this data, security analysts can identify potential threats, such as malicious traffic or unusual network activity patterns that may indicate a security breach. For example, if an organization detects a large amount of traffic between an internal system and a suspicious external IP address, it may indicate a potential attack [34]. Similarly, if network traffic analysis reveals unusual data transfer or communication patterns, it may indicate a data exfiltration attempt.

5.5.5 Malware analysis

This is the process of dissecting and analyzing malicious software (malware) to understand its behavior, functionality, and potential impact on a system or network. This data source is used in attacker profiling to identify and understand the behavior of malicious actors, including the tools and techniques they use to distribute and execute malware. Malware analysis data sources can include information such as the malware's code and behavior, the attacker's command-and-control infrastructure, and the methods used to distribute the malware (for example, email attachments, drive-by downloads, and malicious websites). This information can be gathered through various means, such as malware sandboxing, reverse engineering, and analysis of network traffic associated with the malware [35].

By analyzing malware analysis data, organizations can develop an understanding of the techniques and tools used by attackers to compromise their systems or steal their data. This information can be used to identify potential vulnerabilities in existing security controls and develop new controls that can help prevent or mitigate attacks. For example, suppose malware analysis reveals that a specific type of malware is being distributed via email attachments. In that case, an organization can implement additional email filters and security controls to block these attachments and prevent infection. Similarly, if malware analysis reveals a new type capable of bypassing existing security controls, an organization can develop new controls to address this threat.

5.5.6 Social media monitoring

Social media monitoring is a data source used in attacker profiling to monitor social media platforms for potential threats and malicious activity. Attackers often use social media platforms like Twitter, Facebook, LinkedIn, and others to share information, coordinate attacks, and distribute malware. Social media monitoring involves tracking social media accounts, keywords, hashtags, and other relevant data to identify potential threats and malicious activity. This can include monitoring accounts that are known to be associated with malicious actors, tracking hashtags or keywords that are commonly used concerning cyber-attacks, and monitoring social media activity for IOCs that may indicate an attack is underway [36].

By analyzing social media monitoring data, organizations can understand the types of attacks being planned or executed and the TTPs used by attackers. This information can be used to develop proactive security measures that can help prevent attacks and reactive measures to mitigate the impact of an attack that has already occurred. For example, suppose social media monitoring reveals that a specific group of threat actors is planning a targeted attack on an organization. In that case, the organization can increase its security posture and implement additional security controls to prevent the attack from being successful. Similarly, suppose social media monitoring reveals that a specific type of malware is being distributed via social media. In that case, an organization can update its security controls to prevent this type of attack. In addition, analyzing social media activity can provide insights into the mindset and motivations of potential attackers and their potential targets.

5.5.7 Incident response data

Incident response data is a source of information used to make profiles of attackers. This is done by analyzing the data made during and after a security incident. Incident response data can include information such as the time of the incident, the type of attack, the affected systems and data, and the response actions taken to contain and remediate the incident. By analyzing incident response data, security analysts can develop an understanding of the TTPs used by attackers to compromise their systems, as well as the vulnerabilities that were exploited. This information can be used to improve incident response processes, develop new security controls, and update existing security policies and procedures to better protect against future attacks. Incident response data is valuable for attacker profiling [37]. It provides organizations with important information about past security incidents, allowing them to learn from their mistakes and develop more effective security strategies to protect against future attacks. In addition, analyzing data from previous incidents can help identify patterns and inform response strategies for future attacks.

5.5.8 Insider threat data

Insider threat data is a data source used in attacker profiling to identify and understand the risks posed by internal employees or contractors who may have access to sensitive systems or data. Insider threats can include individuals who intentionally or unintentionally compromise security, such as employees who steal sensitive data, contractors who inadvertently introduce malware onto a network, or employees who accidentally disclose confidential information.

Insider threat data can include information such as employee or contractor access logs, data on employee or contractor behavior, and other relevant data that can help identify potential insider threats. This data can be gathered from various sources, including network logs, security cameras, and other monitoring tools. By analyzing insider threat data, organizations can develop an understanding of the behaviors and patterns that may indicate a potential insider threat [36, 38]. This information can be used to develop policies and procedures that mitigate the risks posed by insiders and identify and investigate potential insider threats before they can cause significant harm. Monitoring internal data can help identify potential insider threats and inform strategies to prevent or mitigate them.

5.5.9 Physical security data

Physical security data is a source used in attacker profiling to identify and understand the risks posed by physical threats to an organization's assets, facilities, and personnel. Physical threats can include theft, vandalism, terrorism, and other forms of physical attack. In addition, physical security data can include access control logs, security camera footage, and other relevant data that can help identify potential physical threats [39]. This data can be gathered from various sources, including security systems, physical security personnel, and other monitoring tools. This information can be used to develop and implement more effective physical security measures to prevent or mitigate the impact of physical attacks. In addition, analyzing data from physical security systems, such as surveillance cameras and access control systems, can help identify potential threats and vulnerabilities.

5.5.10 Human intelligence (HI)

HI data is used in attacker profiling. It involves gathering information from individuals with direct knowledge or experience of an attacker, their motivations, and their methods. This can include information gathered from informants, witnesses, or other sources with first-hand knowledge of the attacker or their activities. HI data can provide valuable insights into the TTPs used by attackers and their motivations and objectives [35]. This information can be used to develop more effective security strategies to detect and prevent future attacks. Additionally, gathering information from human sources, such as informants and whistle-blowers, can provide valuable insights into potential threats and attackers.

5.5.11 Security measures

Implement strong passwords and multi-factor authentication. Strong passwords with a mix of upper- and lower-case letters, numbers, and symbols can be complicated for hackers to crack. Multi-factor authentication provides an additional layer of security by requiring users to provide two or more pieces of evidence to prove their identity. Keep the software up-to-date [36]. Outdated software can be vulnerable to cyber-attacks, as security patches and updates may not have been installed.

Keeping software up-to-date with the latest security patches is essential for protecting against known vulnerabilities. Encryption is converting data into a code to prevent unauthorized access. Using encryption for sensitive data, such as passwords, credit card numbers, and personal information, can help protect against data breaches. Backup data: Regularly backing up data can help prevent data loss in a cyber-attack, such as a ransomware attack. Multiple backups in different locations, such as cloud storage and external hard drives, can provide additional protection.

Implement firewalls: Firewalls can help prevent unauthorized access to networks and systems by blocking incoming traffic from suspicious sources. Train employees: Employee training and awareness are crucial for preventing cyber-attacks. Educating employees on best practices, such as identifying phishing emails and creating strong passwords, can help reduce the risk of cyber incidents [37, 38]. Conduct regular security audits: Regular security audits can help identify vulnerabilities in networks and systems and ensure that security measures are up-to-date and effective.

5.6 Conclusion

Cyber attacker profiling is necessary to gather and analyze various data sources to identify and understand cyber attackers' TTPs. These data sources include attack histories, threat intelligence, vulnerability databases, network traffic data, malware analysis, social media monitoring, incident response data, insider threat data, physical security data, and HI data. Cyber attacker profiling allows organizations to develop more effective security strategies to detect, prevent, and respond to cyber-attacks. Furthermore, by understanding the TTPs used by attackers, organizations can implement more targeted and robust security controls to mitigate cyber-attack risks. On the other hand, cyberbullying is a form of online harassment that can have significant negative impacts on its victims. Cyberbullying can take many forms, including threats, insults, and spreading false information. Therefore, individuals and organizations must be aware of cyberbullying and take steps to prevent and respond to it. Overall, cyber attacker profiling and cyberbullying are two distinct but essential topics in the cybersecurity field. By understanding attackers' tactics and methods and promoting safe and respectful online behaviors, we can all work together to create a safer and more secure online environment.

References

[1] Nguyen, T.N. 'A review of cybercrime,' Journal of Social Review and Development, vol. 2, no. 1, pp. 01–03, 2023.

[2] Shafik, W. 'Cyber security perspectives in public spaces: drone case study,' Handbook of Research on Cybersecurity Risk in Contemporary Business Systems, 2023, IGI Global.

[3] Mijwil, M., and Aljanabi, M. 'Towards artificial intelligence-based cybersecurity: the practices and ChatGPT generated ways to combat cybercrime,' Iraqi Journal for Computer Science and Mathematics, vol. 4, no. 1, pp. 65–70, 2023.

[4] Despotović, A., Parmaković, A., and Miljković, M. 'Cybercrime and cyber security in fintech,' in Digital Transformation of the Financial Industry: Approaches and Applications, Springer, 2023, pp. 255–272.

[5] Mostafavi, S., and Shafik, W. 'Fog computing architectures, privacy and security solutions,' Journal of Communications Technology, Electronics and Computer Science, vol. 24, pp. 1–14, 2019.

[6] Drobotov, S., Pertsev, R., Hrab, M., Fedytnyk, V., Moroz, S., and Kikalishvili, M. 'Forensic research of the computer tools and systems in the fight against cybercrime,' Information Technology and Management, vol. 15, no. 1, pp. 135–162, 2023.

[7] Tok, Y.C., and Chattopadhyay, S. "Identifying threats, cybercrime and digital forensic opportunities in smart city infrastructure via threat modeling," Forensic Science International: Digital Investigation, vol. 45, pp. 301540, 2023.

[8] Verma, S., Sharma, J., Kaushik, K., and Vyas, V. "Mounting cases of cyber-attacks and digital payment," in Cybersecurity Issues, Challenges, and Solutions in the Business World, IGI Global, 2023, pp. 59–80.

[9] Shafik, W., and Matinkhah, S.M. 'Privacy issues in the social web of things,' 5th International Conference on Web Research (ICWR), Tehran, Iran, pp. 208–214, 2019.

[10] Mijwil, M., Salem, I.E., and Ismaeel, M.M. 'The significance of machine learning and deep learning techniques in cybersecurity: a comprehensive review,' Iraqi Journal for Computer Science and Mathematics, vol. 4, no. 1, 87-101, 2023.

[11] Jun, Y., Craig, A., Shafik, W., and Sharif, L. 'Artificial intelligence application in cybersecurity and cyberdefense,' Wireless Communications and Mobile Computing, 1–10, 2021.

[12] Thapaliya, S., and Sharma, P.K. 'Optimized deep neuro fuzzy network for cyber forensic investigation in big data-based IoT infrastructures,'

International Journal of Information Security and Privacy (IJISP), vol. 17, no. 1, 1–22, 2023.
[13] Zhao, L., Zhu, D., Shafik, W., Matinkhah, S.M., and Ahmad, Z. et al., 'Artificial intelligence analysis in cyber domain: A review,' International Journal of Distributed Sensor Networks, vol. 18, no. 4, pp. 15501329221084882, 2022.
[14] Shafik, W., Matinkhah, M., and Sanda, M.N. 'Network resource management drives machine learning: a survey and future research direction,' Journal of Communications Technology, Electronics and Computer Science, 2020.
[15] Winger, G., Ellis, S., and Glover, D. 'Bridging the Digital Gap: Teaching Cyber Strategy and Policy through a Crisis Simulation," International Studies Perspectives, pp. ekad001, 2023.
[16] Shafik, W., Matinkhah, M., Etemadinejad, P., and Sanda, M.N. 'Reinforcement learning rebirth, techniques, challenges, and resolutions,' JOIV: International Journal on Informatics Visualization, vol. 4, no. 3, pp. 127–35, 2020.
[17] Katsantonis, M.N., Manikas, A., Mavridis, I., and Gritzalis, D. "Cyber range design framework for cyber security education and training," International Journal of Information Security, pp. 1–23, 2023.
[18] Chang, V., Golightly, L., Xu, Q.A., Boonmee, T., and Liu, B.S. "Cybersecurity for children: an investigation into the application of social media," Enterprise Information Systems, p. 2188122, 2023.
[19] Shafik, W., Matinkhah, S.M., and Ghasemzadeh, M. 'Internet of things-based energy management, challenges, and solutions in smart cities,' Journal of Communications Technology, Electronics and Computer Science, vol. 27, pp. 1–11, 2020.
[20] Pribadi, B., Rosdiana, S., and Arifin, S. 'Digital forensics on facebook messenger application in an android smartphone based on NIST SP 800-101 R1 to reveal digital crime cases,' Procedia Computer Science, vol. 216, pp. 161–167, 2023.
[21] Yang, Z., Jianjun, L., Faqiri, H., Shafik, W., and Talal Abdulrahman, A. et al., 'Green internet of things and big data application in smart cities development,' Complexity, pp. 1–15, 2021.
[22] Shafik, W., Matinkhah, S.M., Shokoor, F., and Sharif, L. 'A reawakening of machine learning application in unmanned aerial vehicle: future research motivation,' EAI Endorsed Transactions on Internet of Things, vol. 8, no. 29, 2022.
[23] Gaol, F.L., Dermawan, G., Rafif, M.R., Rubianto, R., and Matsuo, T. "The influence of digital financial literacy on interest in investing in

the capital market of millennial generation students," in Inventive Computation and Information Technologies: Proceedings of ICICIT 2022, Springer, 2023, pp. 189–206.

[24] Steinmetz, K.F., Schaefer, B.P., Brewer, C.G., and Kurtz, D.L. "The Role of computer technologies in structuring evidence gathering in cyber-crime investigations: a qualitative analysis," Criminal Justice Review, pp. 07340168231161091, 2023.

[25] Shafik, W., Matinkhah, S.M., and Shokoor, F. 'Recommendation system comparative analysis: internet of things aided networks,' EAI Endorsed Transactions on Internet of Things, vol. 8, no. 29, 2022.

[26] Seaman, J. 'Zero Trust Security Strategies and Guideline,' in Digital Transformation in Policing: The Promise, Perils and Solutions, Springer, 2023, pp. 149–168.

[27] Shafik, W., Matinkhah, S.M., and Ghasemzadeh, M. "A mobile fuzzy sink scheme for wireless sensor network period improvement," 8th Iranian Joint Congress on Fuzzy and intelligent Systems (CFIS), Mashhad, Iran, pp. 211–216, 2020. IEEE.

[28] Johri, A., and Kumar, S. "Exploring customer awareness towards their cyber security in the kingdom of saudi arabia: a study in the era of banking digital transformation," Human Behavior and Emerging Technologies, 2023.

[29] Shokoor, F., Shafik, W., and Matinkhah, S.M. 'Overview of 5G & Beyond Security,' EAI Endorsed Transactions on Internet of Things, vol. 8, no. 30, 2022.

[30] Shafik, W., and Tufail, A. 'Energy optimization analysis on internet of things,' In: Mabrouki, J., Mourade, A., Irshad, A., Chaudhry, S. (eds) Advanced Technology for Smart Environment and Energy. Environmental Science and Engineering, Springer, Cham 2023.

[31] Chen, L., Liu, X., and Tang, H. 'The interactive effects of parental mediation strategies in preventing cyberbullying on social media,' Psychology Research and Behavior Management, vol. 31, pp. 1009–10022.

[32] Tok, Y.C., and Chattopadhyay, S. 'Identifying threats, cybercrime and digital forensic opportunities in smart city Infrastructure via threat modeling,' Forensic Science International: Digital Investigation, vol. 45, pp. 301540, 2023.

[33] Roche, S., Otarra, C., Fell, I., Torres, C.B., and Rees, S. 'Online sexual exploitation of children in the philippines: a scoping review,' Children and Youth Services Review, vol. 9, pp. 106861, 2023.

[34] Shafik, W., and Matinkhah, S.M. 'A study of reinforcement learning on internet of things,' International Conference on Science, Engineering & Technology, ICSET, Pune, Maharashtra, India, 2019.

[35] Matinkhah, S.M., Shafik, W., and Ghasemzadeh, M. 'Emerging artificial intelligence application: reinforcement learning issues on current internet of things,' IEEE 16th International Conference in Information Knowledge and Technology, Tehran, Iran, 2019.
[36] Fonseca, C.C., Moreira, S., and Guedes, I. 'The prevention and control of online consumer fraud,' In Handbook on Crime and Technology, pp. 395–410, 2023. Edward Elgar Publishing.
[37] AL-Hawamleh, A.M. 'Predictions of cybersecurity experts on future cyber-attacks and related cybersecurity measures,' International Journal of Advanced Computer Science and Applications, vol. 14, no. 2, 2023.
[38] Lilhore, U.K., Simaiya, S., Sandhu, J.K., Baliyan, A., and Garg, A. 'EHML: An efficient hybrid machine learning model for cyber threat forecasting in CPS,' In International Conference on Artificial Intelligence and Smart Communication (AISC), pp. 1453–1458, 2023. IEEE.
[39] Shafik, W. "A comprehensive cybersecurity framework for present and future global information technology organizations," In Effective Cybersecurity Operations for Enterprise-Wide Systems, pp. 56–79. IGI Global, 2023.
[40] Aslam, M.M., Tufail, A., Kim, K.H., Apong, R.A., and Raza, M.T. "A Comprehensive Study on Cyber Attacks in Communication Networks in Water Purification and Distribution Plants: Challenges, Vulnerabilities, and Future Prospects," Sensors, 23(18):7999, 2023.

6

Revolutionizing Medical Aid in Outer Space: A Blockchain-enabled Drone Delivery System

J. Alfred Daniel[1], Awais Ahmad[2], C. Chandru Vignesh[3], and M. Newlin Rajkumar[4]

[1]Department of Computer Science and Engineering, Karpagam Academy of Higher Education, India
[2]Information Systems Department, Imam Mohammed Bin Saud University, KSA
[3]School of Computer Science and Engineering, Vellore Institute of Technology, India
[4]Department of Electrical and Electronic Engineering, Anna University Regional Campus, India

Email: 85.alfred@gmail.com; aahmad.marwat@gmail.com; profchandruvignesh@gmail.com; drmnrkaurccbe@gmail.com

Abstract

Delivering medical supplies in space is challenging due to the unique environment and conditions that space presents. Astronauts who work on space missions may face various health risks, such as exposure to radiation, microgravity, and extreme temperatures. Therefore, medical supplies and equipment availability are critical to their health and well-being. This chapter proposes a blockchain-based secure delivery system using drones for medical supplies in outer space. The unique challenges of delivering medical supplies in space are the high cost of transportation and the difficulty of accessing remote locations, which make traditional delivery methods impractical. To address these challenges, the chapter proposes a system that uses blockchain technology for secure and efficient supply chain management and drones for fast and reliable delivery of medical supplies to astronauts. The system employs a decentralized blockchain network that tracks the movement of medical supplies from the manufacturer to the recipient, ensuring transparency and accountability in the supply chain. Using drones for delivery ensures that medical supplies are delivered quickly and safely to their intended destination. The proposed system has the potential to significantly improve the delivery of medical supplies in space and ensure the health and safety of astronauts. Further, the chapter discusses the technical details of the proposed system and evaluates its feasibility and limitations. Our results show that the system effectively provides a secure and efficient supply chain management system and a reliable delivery system for medical supplies in outer space. This chapter contributes to the growing research on using blockchain and drones for medical supply delivery. It presents a novel solution for the unique challenges of delivering medical supplies in outer space.

6.1 Introduction

Space exploration aims to find, study, and comprehend the immense cosmos beyond our planet. It entails sending human and unmanned trips to planetary systems such as Mars, Moon, and beyond, in addition to performing scientific research and experimentation in the weightless environment of space. While space travel offers enormous promise for scientific discoveries and the growth of human understanding, it also presents several obstacles, notably in providing astronauts with medical care [1]. The unusual atmosphere of outer space provides several problems that may negatively influence astronauts' health and well-being. Radiation exposure, which may be substantially greater in space than on Earth's surface, is one of the key obstacles. Prolonged radiation exposure may negatively affect the human body, including an increased risk of cancer, organ damage, and a weakened immune system.

Another difficulty is microgravity, which happens when the Earth's gravitational pull is greatly diminished. In microgravity, body and physiological functions are disrupted, resulting in a loss of muscle and bone, cardiovascular abnormalities, and a weaker immune system. The absence of gravity also alters fluid distribution inside the body, producing upper-body edema and decreased blood flow to the lower limbs [2]. Moreover, the severe temperatures in space, which, depending on the location, might range from extreme cold to intense heat, can harm astronauts' health and the integrity of medical equipment. It is essential for delivering sufficient medical treatment that medications, medical instruments, and biological data retain their stability and efficacy under such high circumstances.

In addition to the difficulties provided by the space environment, medical help in space presents logistical obstacles. Traditional delivery techniques, including resupply trips and cargo spacecraft, may be expensive and time-consuming. Accessing distant regions or extreme circumstances in space may be extremely difficult due to the need for timely medical assistance to astronaut health and safety [3]. To address these obstacles, creative methods are required to assure the availability, dependability, and efficacy of medical help in space. Utilizing cutting-edge technology, including blockchain and drones, has the potential to transform the transportation of medical supplies and improve astronauts' healthcare skills. By integrating distributed ledger for reliable supply chain management and deploying drones for rapid and dependable delivery, a new paradigm for medical assistance delivery in outer space may be realized, safeguarding the health and safety of astronauts throughout space missions. Availability, timeliness, and efficacy of medical help may be negatively impacted by the limits of conventional medical supply delivery techniques for space missions.

Transporting medical supplies into space through resupply operations or cargo spacecraft may be costly and resource-intensive. Launching and sustaining spacecraft increases mission costs significantly, making regular resupply trips economically problematic. Furthermore, the logistics of arranging and conducting resupply trips according to strict timetables may be difficult and time-consuming. Even with regular resupply operations, there may be considerable delays in providing astronauts with medical supplies. Launch dates, travel time, and operational factors might result in lengthy wait times for life-saving medical care. When immediate medical care is needed, time delays may be detrimental in emergencies [4]. In addition, spaceships and settlements have restricted storage capacity, making it difficult to accommodate various medical supplies. Space limitations need careful optimization and medical supply prioritization; certain goods may need more. Similarly, moving and storing heavy or huge medical equipment in a limited place could take time and effort. Traditional techniques rely on production and supply networks based on Earth, which might lead to dependency on Earth-based resources and support. Interruptions or delays in Earth's supply chain directly impact medical supplies' availability in space. The necessity for constant replenishment from Earth restricts self-sufficiency and might pose concerns in the case of unanticipated events [5].

Traditional approaches may need more adaptability to react rapidly to changing medical requirements or crises. Adjusting the supply chain or responding to unanticipated medical needs may be arduous and time-consuming. The failure to swiftly adjust to developing medical settings may jeopardize the efficacy of medical help provided during space missions. Significant dangers are associated with launching spacecraft and cargo, and any mishaps or malfunctions may result in losing vital medical supplies. In the case of malfunctions during launch, transit, or landing, astronauts' safety and the quality of medical assistance might be jeopardized [6]. It is essential to address these restrictions to ensure the availability, dependability, and efficacy of medical help during space missions. Innovative ways that harness cutting-edge technology, such as blockchain and drones, provide the ability to overcome these obstacles and transform the delivery of medical supplies in outer space. Several issues motivate using blockchain technology and drones for efficient and safe medical assistance delivery during space missions.

Blockchain technology enables a decentralized and immutable record that can trace and verify the transfer of medical products from the maker to the receiver. Integrating blockchain makes the supply chain visible and auditable, decreasing the danger of counterfeit or altered materials. This assures the authenticity and integrity of medical help, offering better confidence and protection for astronauts. Every medical supply transaction and movement can be recorded and validated using blockchain. This improved openness for exact supply chain monitoring, assuring accountability at every point. Astronauts, ground control, and stakeholders can easily track medical supplies' source, processing, and delivery, allowing for effective quality control and regulatory requirements.

By offering real-time information on the availability and condition of medical goods, blockchain technology helps simplify inventory management [7]. This allows for improved resource planning, minimizes the danger of supply shortages, and optimizes the distribution of commodities depending on demand. Efficient inventory management assures that astronauts get accessibility to essential medical assistance when they need it, hence avoiding delays and enhancing the overall delivery of healthcare. In space missions, drones provide a rapid and adaptable delivery system for medical supplies. They can travel difficult terrains, reach isolated sites, and swiftly deliver supplies without depending on typical resupply operations. Drones may dramatically shorten delivery times, enabling astronauts to get timely medical care. In addition, their ability to circumvent Earth-based logistics and transmit straight to the desired location improves the overall efficiency and dependability of medical assistance delivery. Integrating blockchain technology with drones may lower medical supply delivery costs and reliance on Earth-based resources. Streamlining supply chain operations and minimizing the frequency of resupply trips, transportation, and logistics expenses may be reduced. In addition, the use of drones avoids the need for bigger, more costly ships devoted purely to cargo delivery, leading to a more sustainable and effective space exploration environment.

Blockchain technology with drones provides a more versatile and responsive method of delivering medical help in space. Blockchain enables the fast incorporation of new manufacturers and distributors into the supply chain, allowing it to adapt to growing medical needs. Drones can swiftly respond to situations and is capable of swiftly responding to situations and delivering essential medical consumables to astronauts. Combining blockchain with drones gives the flexibility to manage changing medical demands and unanticipated events successfully. Utilizing blockchain technology with drones for medical assistance delivery during space missions is motivated by the pursuit of convenience, reliability, accountability, and adaptability. These technologies can potentially transform the distribution of medical supplies, safeguarding the health and safety of astronauts engaged in space exploration.

6.2 Related Works

Earlier research and studies on medical help in space and drone delivery systems have investigated several healthcare and supply chain management areas of healthcare and supply chain management for space missions.

Microgravity-induced and radiation exposure physiological abnormalities and the necessity for prompt medical intervention are a few of the health concerns linked with space missions that have been studied [8]. These obstacles need the existence of reliable and effective medical help delivery networks. Furthermore, optimizing supply chain management (SCM) in space missions to ensure the availability and on-time delivery of medical supplies has been the subject of research. This covers inventory management, demand forecasting, and the effective use of scarce resources. Additionally, traditional resupply flights have been carefully examined to comprehend the logistics, costs, and restrictions of transporting medical resources to space. Researchers have evaluated the schedule, cargo capacity, and frequency of resupply trips to ensure continual supply availability.

Investigations on drone delivery systems in terrestrial contexts have investigated using unmanned aerial vehicles (UAVs) to convey medical supplies in distant or inaccessible regions [9]. These studies have proved the viability and advantages of employing drones for quick, trustworthy, and cost-effective distribution. Further, considerations for applying drone technology to space missions include mobility, cargo capacity, power needs, and steering in microgravity conditions. These investigations have revealed the capability of drones to solve logistical obstacles and expedite the delivery of medical help in space. Additionally, focus has been placed on developing autonomous navigation and control technologies to allow effective space drone delivery. These systems use sophisticated algorithms, sensors, and communication protocols to guarantee drones' safe and accurate navigation in space conditions.

The potential of blockchain technology to improve supply chain management and ensure the privacy and provenance of medical products has been examined [10]. There have been studies on applying blockchain technology to build visible and auditable supply chains, eliminate counterfeiting concerns, and enhance inventory management. Moreover, some researchers have explored integrating blockchain technology with drone deliveries to provide efficient and secure medical assistance in space [11–17]. These systems intend to use both technologies' advantages to handle supply chain management and delivery issues in space missions.

Henceforth, earlier research and studies have provided the groundwork for comprehending the obstacles and potential associated with providing medical treatment in space. They have investigated the possibilities for drone delivery systems, supply chain management tactics, and blockchain technology to transform medical help in space missions. Future improvements in these fields are anticipated to enhance healthcare capabilities and safeguard the well-being of astronauts during space travel.

6.3 Proposed Methodology

By combining blockchain technology and drones, a new delivery system for medical supplies to astronauts in space is being developed to usher in a new era of efficiency and effectiveness in distributing these essential supplies and tracking and managing the transit of medical goods from the maker to the receiver in outer space via a decentralized blockchain network. The technology uses a decentralized blockchain network to monitor and control the transport of medical goods from the maker to the receiver in space. The blockchain guarantees supply chain transparency and security by recording and validating every transaction and supply movement. This allows stakeholders to track medical assistance's origin, management, and delivery, guaranteeing the supplies' integrity and legitimacy. Using blockchain technology, medicinal supplies may be securely authenticated and verified. A unique digital identification or cryptographic tag may be issued to each supply item and stored on the blockchain.

This minimizes the danger of counterfeit or altered items by ensuring that only authentic and approved materials are involved in the supply chain. Blockchain-enabled technology allows real-time visibility and monitoring of medical supply inventories. The distributed ledger is constantly updated as supplies are added or used, providing precise inventory management. This optimizes resource planning, reduces the chance of shortages, and ensures crucial medical assistance is available when required. In outer space, medical supplies are delivered by unmanned aerial vehicles. These drones are built and outfitted for operation in space, considering agility, cargo capacity, and power needs. They can handle space obstacles such as microgravity, severe temperatures, and isolated areas to deliver medical supplies promptly and reliably. The system's drones are equipped with autonomous guidance and control technologies. These systems use algorithms, sensors, and interoperability to guarantee safe and accurate space navigation [18][19].

The ability of drones to independently plan their routes, avoid barriers, and adjust to changing environmental circumstances guarantees the efficient and dependable delivery of medical supplies. The blockchain-enabled system employs encrypted communication protocols to allow communication between the public blockchain, ground stations, and drones. This guarantees that the delivery process stays secure and essential data, like supply chain upgrades and instructions, are transferred securely. The suggested system intends to create a safe, efficient, and dependable delivery mechanism for medical help in outer space by combining the advantages of blockchain technology with unmanned aerial vehicles. It provides openness and accountability of the supply chain, optimizes inventory management, and allows the exact and timely delivery of medical supplies to astronauts on space missions.

The proposed blockchain-enabled drone delivery system for medical supplies in outer space comprises several important components that collaborate to guarantee the effective and safe delivery of medical supplies. Here are the essential elements:

6.3.1 Blockchain framework

The blockchain architecture serves as the supply chain management system's base. It may be done via blockchain platforms like Ethereum, Hyperledger Fabric, or bespoke solutions. The architecture consists of a decentralized network of nodes that verify and log transactions associated with the flow of medicinal goods. It guarantees supply chain transparency, immutability, and consensus, allowing for secure monitoring and accountability.

6.3.2 Drone design

The drone design is vital for delivering medical supplies in deep space. Drones must be properly developed in space conditions, considering microgravity, severe temperatures, and isolated places. The mobility, power needs, cargo capacity, and navigation characteristics are crucial design factors. Drones should be equipped with sensors, communications networks, and autonomous navigation systems to guarantee safe and effective delivery.

6.3.3 Communication protocols

Secure communication methods are required for blockchain networks, mission control, and drone communication. These protocols guarantee the safe and reliable communication of vital data, like supply chain management, delivery instructions, and authentication mechanisms. Encryption, authentication, and data quality safeguards should be applied to prevent unwanted access or manipulation of the communication channels.

6.3.4 Payload handling

Payload handling refers to how drones securely transport and deliver medical supplies. It involves designing specialized containers or compartments within the drones to accommodate and protect medical supplies during transportation. Payload handling relates to how drones safely carry and distribute medical goods. It includes the creation of specific containers or compartments inside drones to secure and convey medical goods.

Key components of the system include the blockchain architecture for the supply chain, drone design for deep space missions, networking devices for secure data transfer, and payload handling procedures to guarantee the secure transportation and distribution of medical goods. Together, these components provide an efficient and safe system for medical help in space.

6.4 Proposed Blockchain-enabled Drone Delivery System (BEDDS) Architecture

The blockchain architecture serves as the supply chain management system's base. It may be done via blockchain platforms like Ethereum, Hyperledger Fabric, or bespoke solutions. The architecture consists of a decentralized network of nodes that verify and log transactions associated with the flow of medicinal goods. It guarantees supply chain transparency, immutability, and consensus, allowing for secure monitoring and accountability. Several important components of the proposed architecture for a blockchain-enabled drone delivery system for medical supplies in outer space assure safe and efficient supply chain management and dependable delivery of medical supplies

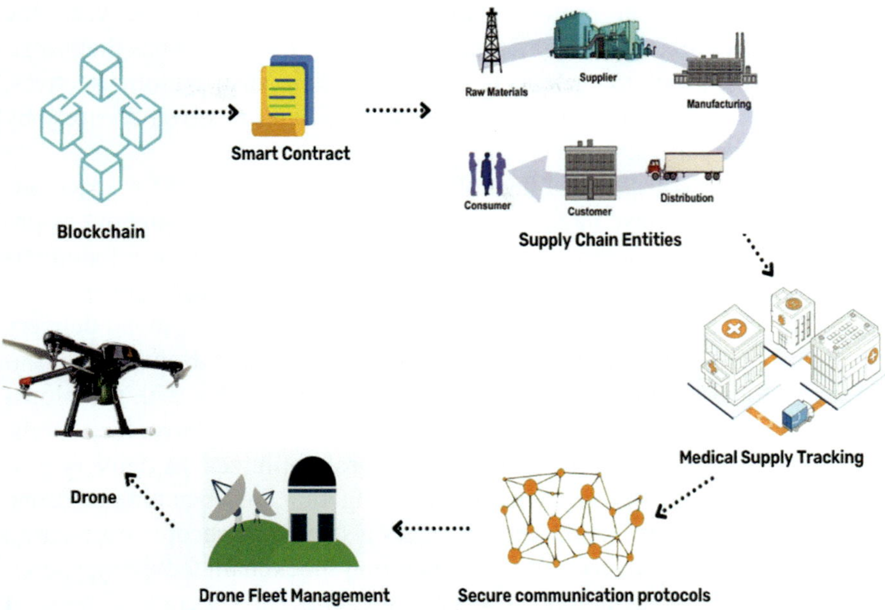

Figure 6.1 Proposed blockchain-enabled drone delivery system architecture.

as illustrated in Figure 6.1. For safe and accessible supply chain management, this solution leverages a decentralized blockchain network. Multiple nodes within the blockchain network verify and record transactions about the transportation of medical supplies. These contracts provide rules and logic that regulate the flow of medical supplies, such as identification, verification, and tracking. Smart contracts allow the automated and tamper-proof execution of supply chain procedures, assuring efficiency and transparency.

These contracts contain predefined rules and logic that govern the flow of medical supplies, including authentication, verification, and tracking. Smart contracts enable automated and tamper-proof execution of supply chain operations, ensuring transparency and efficiency. The system includes manufacturers, suppliers, space agencies, launch control centers, and astronauts, among its supply chain players. Each member in the blockchain network has a specified role and access rights, enabling them to engage with supply chain data depending on their permission level. Unique digital IDs or cryptographic tags are issued to medical supplies and stored on the blockchain. These identities allow reliable supply chain monitoring and traceability. The flow of goods, from manufacture to distribution, is documented on the blockchain as transactions, assuring accountability and transparency.

Secure communication protocols facilitate communication between the blockchain network, ground stations, and drones. This guarantees that sensitive information, such as changes to the supply chain, distribution directives, and authentication information, is communicated and received securely by the appropriate parties.

The system consists of a fleet of drones built particularly for space settings. These drones have navigation systems, sensors, and throughput essential for transporting medical supplies. They are competent in autonomous navigation, route optimization, collision avoidance, and circumstances adaptation. The drones communicate with the blockchain network to get delivery orders and update supply chain data. Ground control is the drone fleet's control tower for oversight and management. It communicates with the blockchain network to access real-time supply chain data, track inventory levels, and coordinate drone operations. Ground control optimizes the delivery process by ensuring the optimal scheduling, navigation, and planning of drone deliveries. The suggested architecture offers a stable and secure space-based medical assistance delivery system by merging blockchain technology, consensus mechanisms, symmetric encryption, drones, and a control tower. It offers responsible and transparent supply chain management, effective inventory monitoring, and dependable drone delivery, safeguarding the health and safety of astronauts throughout space missions.

6.5 Workflow

Several processes are included in the proposed blockchain-enabled drone delivery method for medical help in outer space to ensure a fluid and transparent workflow. Whenever astronauts or ground control submit a request for medical supplies, the procedure begins. The request outlines the amount and location of the medical supplies in outer space as mentioned in Figure 6.2. The request is validated to ensure it conforms to the approved medical supply list and inventory data. The blockchain network's smart contracts validate the request's legitimacy and compliance with preset rules and conditions. The request is stored on the blockchain after confirmation, producing a transparent and unchangeable record of the supply chain activity. This record contains the request timestamp, the requester's details, the requested supplies, and the delivery destination.

In the proposed model, tracking the flow of medical products from the producer to the delivery location, the blockchain network helps supply chain management. The blockchain network's smart contracts validate the request's legitimacy and compliance with preset rules and conditions. The request is

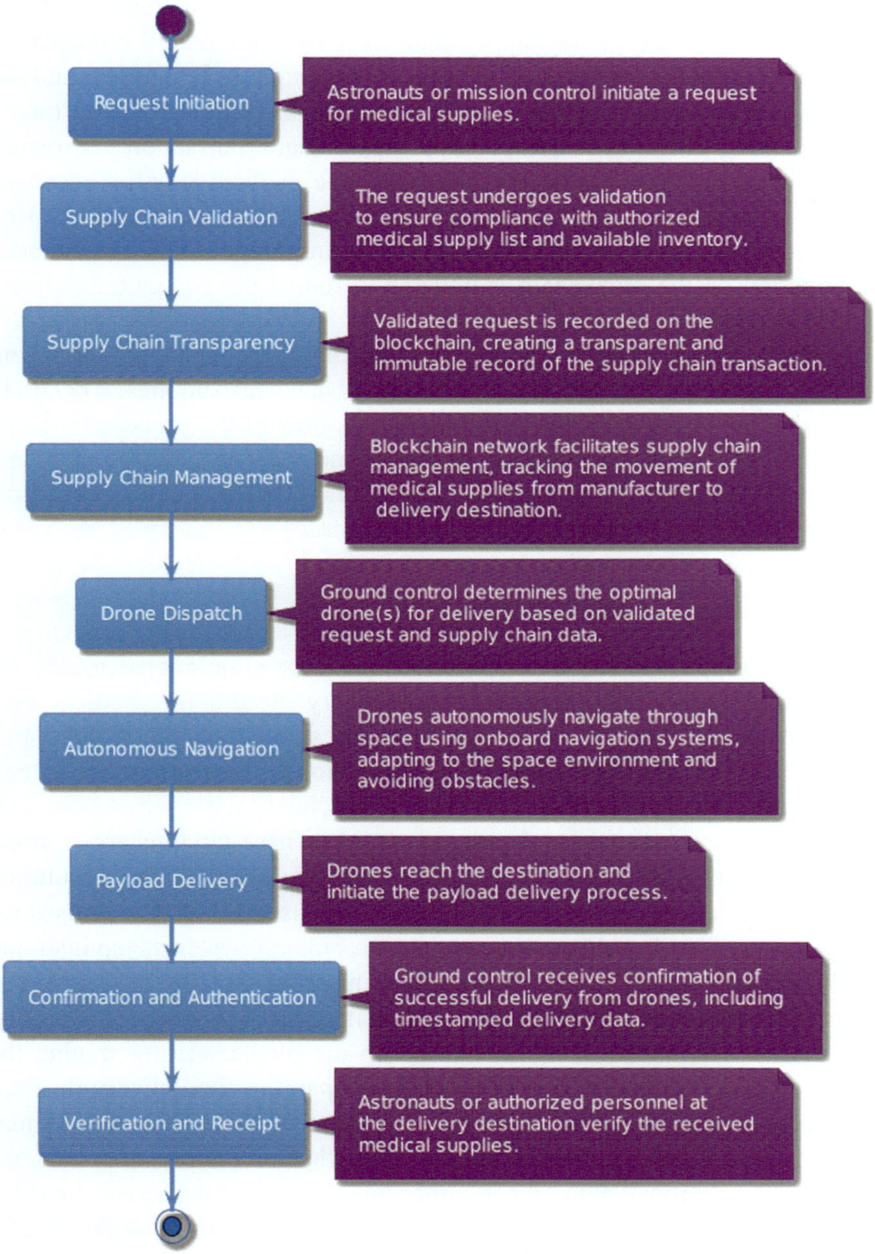

Figure 6.2 Workflow of the proposed blockchain-enabled drone delivery system.

stored on the blockchain after confirmation, producing a transparent and unchangeable record of the supply chain activity. This record contains the request timestamp, the requester's details, the requested supplies, and the delivery destination. The delivery confirmation is recorded on the blockchain, linking it with the corresponding supply chain transaction. Astronauts or authorized staff at the destination check the medical supplies received. Every verification, including any anomalies or problems, is recorded on the blockchain, assuring accountability and transparency. The supply chain information on the blockchain is updated to reflect the completed delivery and current inventory levels.

This upgrade enables real-time monitoring of the supply chain, inventory management, and precise monitoring of medical supplies. The listed phases demonstrate the basic delivery process flow of the proposed system. Incorporating blockchain technology, autonomous drones, and supply chain management enables transparency, security, and efficiency across the whole process, improving medical help delivery in space.

6.6 Results and Discussion

Several assumptions may be made to establish the simulation scenario for a proposed blockchain-enabled drone delivery system for medical help in outer space. These assumptions assist in establishing the simulation's parameters and limitations and give a foundation for analyzing the system. The simulation considers microgravity's special space environment features, severe temperature, and radiation levels. These circumstances affect the performance and behavior of drones and the system. The simulation implies that the system's drones are created exclusively for space missions. They are equipped with the appropriate propulsion systems, navigational skills, cargo handling devices, and communication systems to function in space. The simulation assumes a set of specified medical supplies, quantity, and system availability. The data stored on the blockchain for tracing the transit of goods from the manufacturing company to the delivery destination is expected to be correct. The simulation presupposes using a particular blockchain framework, like Ethereum, Hyperledger Fabric, or a bespoke blockchain solution. Among the assumptions may be the consensus method, block size, transaction validation duration, and any other framework-specific factors.

Table 6.1 tabulates the blockchain-enabled drone delivery system's results over time. The time represents the implementation duration, and subsequent columns display the corresponding metrics such as delivery time,

Table 6.1 Blockchain-enabled drone delivery system's results over time.

Time (months)	Delivery time (hours)	Supply chain efficiency (%)	Cost savings (%)
0	24	85	0
3	16	90	15
6	10	92	25
9	8	94	30
12	6	96	35

Figure 6.3 Supply chain efficiency vs. cost savings.

supply chain efficiency, and cost savings as shown in Figure 6.3. Furthermore, delivery time represents when medical supplies are delivered in hours as shown in Figure 6.4. The delivery time decreases as the system is optimized and improved over time. Moreover, the supply chain efficiency indicates the efficiency of the supply chain management system implemented with blockchain technology. Higher percentages indicate improved efficiency and smoother operations. Additionally, cost savings represents the percentage of cost savings achieved by using the blockchain-enabled drone delivery system compared to traditional methods. As the proposed system becomes more optimized and streamlined, cost savings increase.

The simulation implies that the blockchain network, control tower, and drones can access a dependable communication infrastructure. The

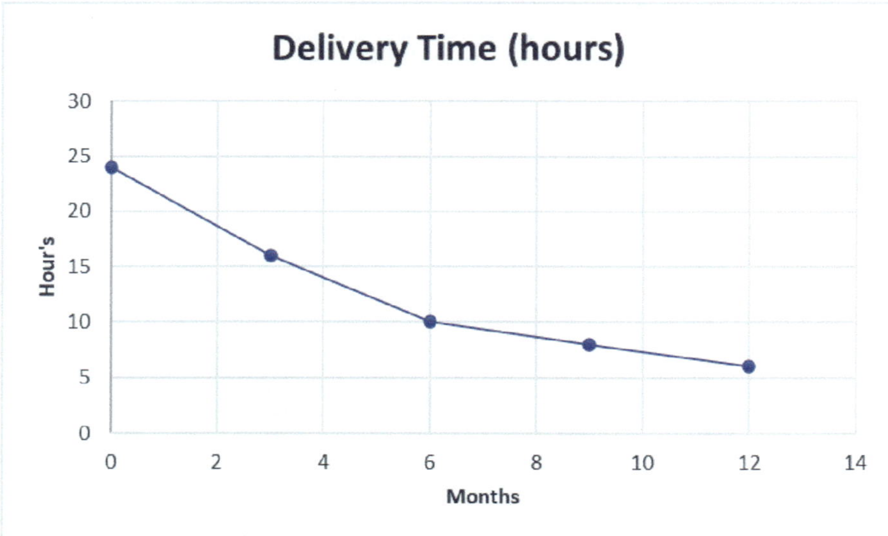

Figure 6.4 Tentative delivery time estimation.

considerations could include communication routes' range, throughput, and latency. The simulation implies a procedure for starting requests for medical supplies, verifying them, deploying the relevant drones, and carrying out the delivery. The predictions may include the time necessary for request approval, drone deployment, navigation systems, and payload distribution. To assess the system's success, the simulation assumes certain performance indicators. These could include delivery schedule, dependability, supply chain visibility, privacy, communication efficacy, and other pertinent indicators, as described earlier. Describing and precisely characterizing the simulation's underlying assumptions is essential to maintain openness and repeatability. The framework of the proposed system's goals, needs, and restrictions, these assumptions may be modified or revised. The interpretation of data acquired from analyzing the performance and efficiency of a blockchain-enabled drone delivery system for medical help in space may have important consequences for space exploration and medical aid in space missions.

6.7 Limitations or Challenges

Implementing a complicated system composed of blockchain technology, drones, and communication systems might bring technological difficulties. Integration concerns, compatibility issues, and hardware limits of drones or blockchain technologies may need to be resolved to guarantee the seamless running of the system. Space's microgravity, harsh temperatures, and radiation pose hurdles to drone functioning and hardware durability. There may be a need for modifications and changes to guarantee that drones can operate successfully and securely in these situations. Utilizing drones in outer space requires compliance with certain rules and safety standards. Compliance with international space regulations, collaboration with space organizations, and guaranteeing the system does not disrupt other space operations are crucial factors. Establishing reliable and effective communication between the blockchain system, ground stations, and space drones may be difficult. Handling data transmission delays and maintaining continual connections in faraway regions must be addressed. It is necessary to ensure the safe and secure transit and delivery of medical goods. The system should be built to accommodate delicate or temperature-sensitive payloads and protect against possible dangers, like collisions or contamination.

Power and energy management optimization is essential for the endurance of space-based drone operations. The framework should be developed to reduce energy use, investigate alternate power sources, and consider recharging or reloading techniques. Real-world testing in space-like circumstances may be difficult and expensive. Simulations and ground-based testing may be useful, but validating the system's performance in a setting that closely reflects its real working settings is essential. Implementing and maintaining a blockchain-enabled drone delivery system in space might be costly. Cost issues, scalability, and financial viability must be extensively analyzed to assure the system's practicability and long-term viability. Managing these constraints and obstacles involves careful planning, cooperation with specialists from various sectors, iterative development, and continual improvement. In order to overcome these obstacles and guarantee the effective deployment of the blockchain-enabled drone delivery mechanism for medical help in space, it is vital to work closely with space agencies, regulatory organizations, and other stakeholders.

6.8 Future Enhancement

The blockchain-enabled drone delivery system for medical help in outer space offers enormous growth and improvement potential. Drones can operate more effectively and securely in complicated space settings if their autonomous navigation skills are improved. Advanced algorithms and sensors may offer real-time obstacle recognition and avoidance, assuring the transportation of medical supplies without collisions. Integrating AI with machine learning approaches may improve the system's decision-making skills. AI algorithms may evaluate data from various sources, such as drone telemetry, blockchain transactions, and medical supply inventories, to optimize delivery routes, estimate supply demand, and identify abnormalities or systemic concerns. Exploring the notion of swarm robotics, where several drones cooperate and coordinate their operations, might improve the system's scalability and robustness. Swarm robots can increase delivery efficiency, allow redundant operations, and react to dynamic settings, providing ongoing medical assistance delivery even under difficult conditions. Utilizing edge computing and onboard processing capabilities in drones helps expedite data analysis and decision-making. By processing data locally, the system may decrease latency, improve reaction times, and decrease dependency on ground-based infrastructure for real-time decision-making.

Exploring the interoperability across various blockchain networks helps increase the system's adaptability and compatibility. Interoperability protocols allow smooth connection with current blockchain platforms used by space agencies or medical institutions, fostering data exchange and assuring interoperability with larger supply chain ecosystems. Developing specialized payload handling mechanisms and containers that can safely transfer many medical supplies, including delicate or perishable products, may expand the system's capabilities. This comprises temperature control systems, shock-absorbing packaging, and secure locking mechanisms. Integrating the drone delivery system with telemedicine capabilities enables astronauts to get remote medical consultations and advice. Drones may deliver medical gadgets that support telemedicine consultations, enabling astronauts to obtain professional medical advice and help in real time. Implementing real-time monitoring of drone health, blockchain network performance, and supply chain data enables proactive maintenance and identifies possible problems in advance of interruptions. Predictive analytics can improve maintenance schedules, decrease downtime, and guarantee that the system performs well.

Streamlining the supply chain and ensuring seamless integration with the broader space ecosystem may be achieved by coordinating the drone delivery system with space habitats and infrastructure, such as space stations and lunar bases. This might include creating specific docking stations, adding robotic arms for cargo transfer, or integrating the system with existing logistical infrastructure. These advancements and improvements have the potential to further transform the delivery of medical help in space, making it more efficient, dependable, and adaptable to the difficulties of space travel. Continuous research, innovation, and cooperation across different disciplines will drive the system's growth and pave the road for future missions with sophisticated healthcare capabilities.

6.9 Conclusion

The chapter proposes a blockchain-enabled drone delivery system for extraterrestrial medical assistance. The report addresses the specific obstacles of delivering medical supplies in space, such as high transportation costs and reaching distant regions, which render conventional distribution techniques impracticable. The chapter illustrates the advantages of using blockchain technology for safe and efficient supply chain management and drones for the rapid and dependable delivery of medical goods into space. These technologies solve the delivery process's concerns of cost, accessibility, and accountability. The chapter illustrates the advantages of using blockchain technology for safe and efficient supply chain management and drones for the rapid and dependable delivery of medical goods into space. These technologies solve the delivery process's concerns of cost, accessibility, and accountability. The chapter details the delivery procedure, from request start through final delivery confirmation. It stresses using blockchain technology to monitor the movement of medical supplies and drones for effective and timely delivery. The chapter outlines the simulation tests undertaken to assess the performance and efficiency of the proposed system. The results illustrate the system's efficiency in providing safe supply chain management and dependable delivery of medical goods in space. The chapter addresses the constraints and obstacles observed throughout the installation and testing stages, emphasizing the necessity for revisions and future advances. It also investigates future improvements, like autonomous navigation, AI integration, and swarm robots. In conclusion, the research study thoroughly examines the blockchain-enabled drone delivery system for medical help in outer space, stressing its advantages, implementation issues, and future improvements. The results demonstrate the practicability and possible influence of the suggested system on space exploration and medical assistance distribution.

References

[1] Dong, Chengzu, Jingwen Zhou, Qi An, Frank Jiang, Shiping Chen, Lei Pan, and Xiao Liu. 2023. "Optimizing Performance in Federated Person Re-Identification through Benchmark Evaluation for Blockchain-Integrated Smart UAV Delivery Systems" Drones 7, no. 7: 413. https://doi.org/10.3390/drones7070413

[2] Feng, C., Liu, B., Yu, K., Goudos, S.K., and Wan, S. "Blockchain-empowered decentralized horizontal federated learning for 5G-enabled

UAVs". IEEE Trans. Ind. Inform. 2021, 18, 3582–3592. Gartner Report, Financial Times, 2007.

[3] Khan,M.A.,Ullah,I.,Alkhalifah,A.,Rehman,S.U.,Shah,J.A.,Uddin,M.I., Alsharif, M.H., and Algarni, F. A provable and privacy-preserving authentication scheme for UAV-enabled intelligent transportation systems. IEEE Trans. Ind. Inform. 2021, 18, 3416–3425.

[4] Pokhrel, S.R., and Choi, J. Federated Learning With Blockchain for Autonomous Vehicles: Analysis and Design Challenges. IEEE Trans. Commun. 2020, 68, 4734–4746.

[5] Gupta, R., Bhattacharya, P., Tanwar, S., Kumar N., and Zeadally, S. "GaRuDa: A Blockchain-Based Delivery Scheme Using Drones for Healthcare 5.0 Applications," in IEEE Internet of Things Magazine, vol. 4, no. 4, pp. 60-66, December 2021, doi: 10.1109/IOTM.001.2100045.

[6] Ali, M.H., Jaber, M.M., Alfred Daniel, J., Vignesh, C.C., Meenakshisundaram, I., Kumar, B. S., and Punitha, P. (2023). Autonomous vehicles decision-making enhancement using self-determination theory and mixed-precision neural networks. Multimedia Tools and Applications, 1–24.

[7] Khan,M.A.,Ullah,I.,Alkhalifah,A.,Rehman,S.U.,Shah,J.A.,Uddin,M.I.; Alsharif, M.H., and Algarni, F. A provable and privacy-preserving authentication scheme for UAV-enabled intelligent transportation systems. IEEE Trans. Ind. Inform. 2021, 18, 3416–3425. Gartner Report, Financial Times, 2007.

[8] Yazdinejad, A., Parizi, R.M., Dehghantanha, A., and Karimipour, H. Federated learning for drone authentication. Hoc Netw. 2021, 120, 102574.

[9] John William, A. D., Rajendran, S., Pranam, P., Berry, Y., Sreedharan, A., Gul, J., and Paul, A. (2022). Blockchain Technologies: Smart Contracts for Consumer Electronics Data Sharing and Secure Payment. Electronics, 12(1), 208.

[10] Velusamy, P., Rajendran, S., Mahendran, R.K., Naseer, S., Shafiq, M., and Choi, J.G. (2021). Unmanned Aerial Vehicles (UAV) in precision agriculture: Applications and challenges. Energies, 15(1), 217.

[11] Rao, P.V.V., Anand, M., and Daniel, J.A. et al. Millimeter assisted wave technologies in 6G assisted wireless communication systems: a new paradigm for 6G collaborative learning. Wireless Netw (2023). https://doi.org/10.1007/s11276-023-03324-6

[12] Zhang, C., Liu, X., Xu, J., Chen, T., Li, G., Jiang, F., and Li, X. An Edge Based Federated Learning Framework for Person Re-Identification in UAV Delivery Service. In Proceedings of the 2021 IEEE International

Conference on Web Services (ICWS), Chicago, IL, USA, 5–10 September 2021; pp. 500–505.

[13] Singh, M., Aujla, G.S., and Bali, R.S. et al. CovaDel: a blockchain-enabled secure and QoS-aware drone delivery framework for COVID-like pandemics. Computing 104, 1589–1613 (2022).

[14] Tri Nguyen, Risto Katila, and Tuan Nguyen Gia. An advanced Internet-of-Drones System with Blockchain for improving quality of service of Search and Rescue: A feasibility study, Future Generation Computer Systems, Volume 140, 2023, Pages 36-52, ISSN 0167-739X, https://doi.org/10.1016/j.future.2022.10.002.

[15] Kim, J., Kim, S., Jeong, J., Kim, H., Park, J., and Kim, T. "CBDN: Cloud-Based Drone Navigation for Efficient Battery Charging in Drone Networks," in IEEE Transactions on Intelligent Transportation Systems, vol. 20, no. 11, pp. 4174–4191, Nov. 2019, doi: 10.1109/TITS.2018.2883058.

[16] Abdellatif, T., Sedrine, M.A., and Gacha, Y. "DroMOD: A Drone-Based Multi-Scope Object Detection System," in IEEE Access, vol. 11, pp. 26652–26666, 2023, doi: 10.1109/ACCESS.2023.3253767.

[17] Silalahi, S., Ahmad, T., and Studiawan, H. "Transformer-Based Named Entity Recognition on Drone Flight Logs to Support Forensic Investigation," in IEEE Access, vol. 11, pp. 3257–3274, 2023, doi: 10.1109/ACCESS.2023.3234605.

[18] Sathish Kumar, Sarveshwaran Velliangiri, Periyasami Karthikeyan, Saru Kumari, Sachin Kumar, and Muhammad Khurram Khan. A survey on the blockchain techniques for the Internet of Vehicles security, Transactions on Emerging Telecommunications Technologies

[19] Velliangiri, S., and Karthikeyan, P. "Blockchain Technology: Challenges and Security issues in Consensus algorithm," 2020 International Conference on Computer Communication and Informatics (ICCCI), Coimbatore, India, 2020, pp. 1–8.

7

A Secure Resilient Scheme for Autonomous Vehicles against External Attacks

J. Jenefa, E. A. Mary Anita, Rakoth Kandan Sambandam, Divya Vetriveeran, and D. Vinodha

Department of CSE, CHRIST (Deemed to be University), India

Email: jenefa.j@christuniversity.in; maryanita.ea@christuniversity.in; rakoth.kandan@christuniversity.in; divya.vetriveeran@christunivrsity.in; vinodha.d@christuniversity.in

Abstract

Autonomous vehicular ad hoc networks are networks created with autonomous vehicles and other entities in the vehicular environment. Like traditional vehicular ad hoc networks, autonomous ad hoc networks are also prone to internal and external attacks. Many authentication schemes are proposed to overcome internal attacks, whereas external attacks are not focused on. Though the impact of external attacks is less when compared to that of internal attacks, external attackers observe and analyze the network traffic information, which will be helpful for the internal attackers to affect the performance of the network. Hence, this chapter proposes a secure identity-based authentication scheme without pairings against external attacks. It uses an elliptic curve cryptography–based identity-based signature to authenticate vehicles. The proposed authentication scheme ensures secure vehicular communications, including inter-vehicular communication, without RSUs during emergencies. Simulation results demonstrate its superior performance.

7.1 Introduction

Autonomous vehicles (AVs) are self-driving vehicles in auto-pilot mode. Without human intervention, they can travel from source to destination

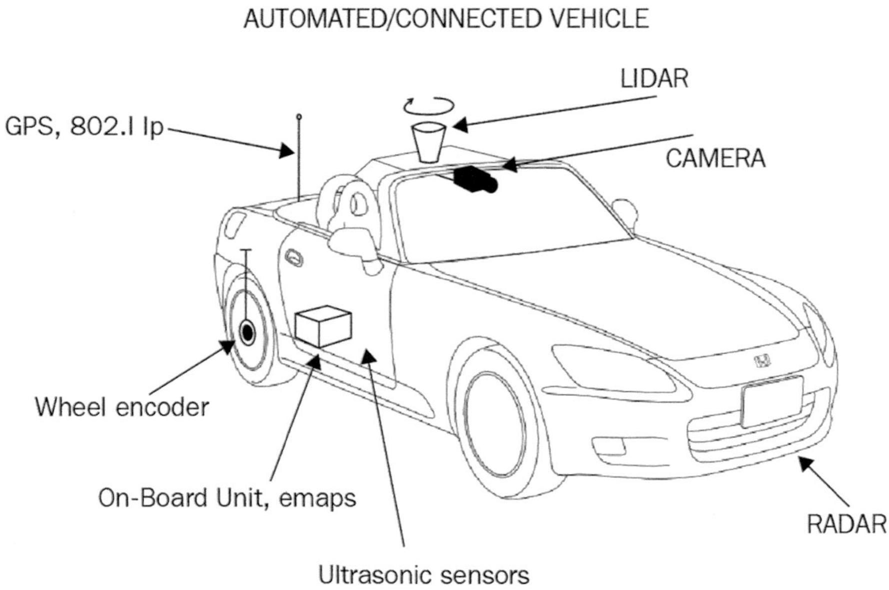

Figure 7.1 Components in autonomous vehicles [4].

independently [1]. This is possible with the help of in-built components like sensors, actuators, perception, motion control components, etc. Today, most famous vehicle manufacturers [2] are deploying semi-automatic driving features like automatic braking, crash avoidance, lane-changing assistance, automatic cruise control, etc., in their vehicles [3]. With technological advancement, it is assumed that autonomous vehicles will rule out traditional vehicles with many additional assistance features.

On the other hand, it also reduces road accidents to a greater extent. Though it has more benefits, deploying a large-scale, fully automated vehicle has various challenges. The components of the autonomous vehicle are given in Figure 7.1.

Autonomous vehicles are based on self-intelligence, and they are based on a single vehicle. Each vehicle observes the environment using different components and performs its actions. With the dynamic nature of the vehicular environment, relying only on self-intelligence may lead to severe consequences. It is also prone to various security attacks. Hence, autonomous vehicles and other entities must communicate. This can be achieved by a network of autonomous vehicles, facilitating communication between the vehicles and other environmental commodities. The communication between the autonomous vehicles is depicted in Figure 7.2. As shown, with the help of

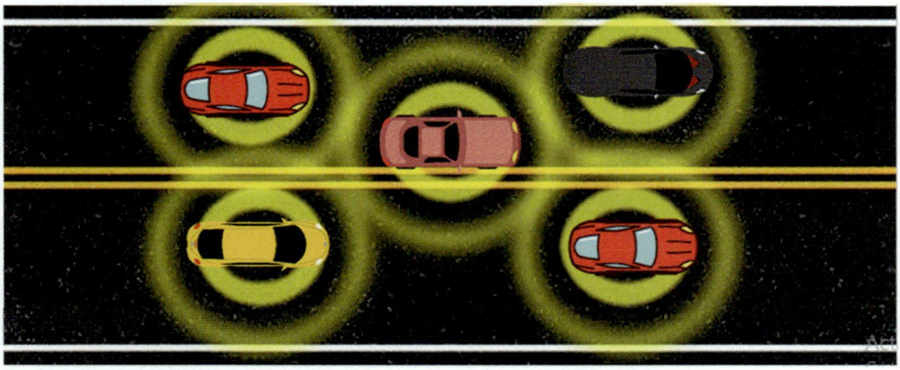

Figure 7.2 Communication in autonomous vehicles.

sensors and other components, AVs can communicate with nearby vehicles in the environment.

Vehicular ad hoc network (VANET) considers moving cars and ensures communications [5]. It has vehicles, trusted authority (TA), and road side units (RSUs) [6]. TAs are allocated to a specific area that manages all the RSUs in that region. RSUs, on the other hand, act as service providers (SPs) for vehicles that provide internet-based services like multimedia services. Additionally, it sends messages about traffic flow to all passing vehicles. Vehicle manufacturers install on-board units (OBUs) to enable communication between the cars. It can use Wi-Fi, WiMax, and other networks to communicate with vehicles and RSUs. An enhanced version of the vehicular ad hoc network with autonomous vehicles is called autonomous vehicular ad hoc networks (AVNETs).

Since autonomous vehicles are the future of vehicular technology, a wide range of research focuses on autonomous vehicular ad hoc networks (AVNETs) [7] and the changes to be incorporated into the existing VANET. The traditional VANET is based on a structural system without intelligence. To make it suitable for the network with autonomous vehicles, different features of the AVs must be considered. The main feature is collaborative driving, which considers multi-car intelligence to ensure a safe journey. Multi-car intelligence can be achieved by sharing the intelligence among the AVs, which is possible through secure vehicular communications. As shown in Figure 7.2, secure communication is ensured in autonomous vehicles. Secure communication is critical to avoiding accidents, and passengers' safety can be guaranteed.

Vehicular communication consists of two types: communication with RSUs and vehicular communications. In communication with RSUs, vehicles request internet-based services from RSUs and send traffic-related messages; vehicles receive traffic-related messages and access services from RSUs [8]. However, both VANETs and AVNETs are vulnerable to security attacks, necessitating establishing secure communication. An efficient authentication scheme is employed to achieve secure communication, enabling vehicles to prove their legitimacy and identify potential attackers in the network [9]. By ensuring that services are provided only to legitimate vehicles, the forwarding of false event messages from attackers throughout the network can be prevented. Several approaches have been put forth to ensure a resilient network.

Security requirements include message integrity, privacy-preservation, confidentiality, and nonrepudiation [10]. Privacy-preservation [11] helps in protecting the original details of vehicles. It evades other network nodes from misusing a vehicle's details. Message integrity ensures that any other attackers in the network do not alter the messages received. Confidentiality helps to reveal the transmitted message only to the intended receiver. Attackers cannot view the message that is transmitted from the sender to the receiver. Nonrepudiation makes the vehicles take responsibility for their actions throughout the network. Vehicles cannot refuse their actions after committing violence.

Identity-based techniques have been put up as a way to create secure communication, with techniques like those in [12], [13], [14], and [15]–[22] gaining attention. These schemes utilize the vehicle's identity for generating signatures and employ ID–based signatures for verifying other vehicles' identities, eliminating the need for certificates. Two categories of IBS schemes are based on pairing operations: those with pairings and those without pairings. Numerous innovative schemes have been proposed within each category. However, it is essential to note that most schemes do not provide adequate security against external attacks.

External attacks pose a challenge as they involve eavesdropping on transmitted messages without altering them or masquerading as other vehicles. These attackers, typically outsiders, collect traffic information throughout the network and conduct traffic analysis. Since they do not engage in any violations and merely observe the traffic for analysis purposes, it becomes difficult to detect them. This chapter proposes a secure IBS-based authentication without pairings, and our contributions are listed below.

7.1.1 Our contributions

The proposed authentication scheme for AVNET offers a secure solution without complex pairing operations, effectively reducing overheads. It provides robust security against external attacks, eavesdropping, man-in-the-middle attacks, and other internal threats. In addition to addressing various attacks, the scheme ensures the fulfillment of fundamental requirements for security.

The rest of the chapter is divided into the following sections:

Section 7.2: Comprehensive review of relevant works.

Section 7.3: The architectural design of the proposed methodology and outlines the ensured essential requirements in security aspects.

Section 7.4: Detailed explanation of the secure authentication scheme proposed in the chapter.

Sections 7.5: Security analysis.

Section 7.6: Performance analysis.

Section 7.7: Conclusion.

7.2 Related Works

Identity-based authentication schemes can be categorized as those with and without pairings. Pairing-based schemes involve complex mathematical operations, while non-pairing schemes aim to reduce computational overhead. These two categories' schemes are explained briefly in this section.

7.2.1 Identity-based authentication scheme with pairings

In [23], a novel authentication framework that employs bilinear pairings is proposed. It utilizes an identity-based signature (IBS) for communication with RSUs and an ID-based online/offline signature (IBOOS) for vehicular communication. It needs to determine how traffic-related messages are communicated after the verification cycle, which is a drawback. The MADAR scheme, proposed in [24], offers authentication with conditional privacy-preservation and resistance to Denial of Service (DoS) attacks. It utilizes bilinear pairings based on IBS and IBOOS schemes. It also ensures nonrepudiation and traceability. Similar to the previous scheme, it shares the limitation of not specifying the transmission of traffic-related messages.

In [25], an ID-based batch verification scheme is presented, employing an elliptic curve cryptography (ECC) based IBS scheme with bilinear pairings. Vehicles create pseudo IDs by leveraging the public parameters of a trusted authority (TA). The scheme provides nonrepudiation, traceability, unlinkability, and conditional privacy-preservation alongside authentication. However, since traffic-related messages are transmitted directly to other vehicles, it is open to external attacks.

The distributed aggregate IBS proposed in [26] uses an ECC-based IBS scheme. Instead of a single TA, it incorporates multiple TAs, and vehicles generate one-time use private keys using the TAs' public parameters. The scheme allows the compression and storage of multiple signatures as a single one. It offers conditional unlinkability but shares the drawback of not addressing the transmission of traffic-related messages.

In [27], a pseudonymous identity-based authentication scheme is introduced, utilizing ECC with bilinear pairings. Vehicles send their identity to the motor vehicles division (MVD) or TA, which authenticates them and provides tickets for generating pseudo IDs with the assistance of an RSU. Sun et al. [28] offer a defense method for vehicle networks that protect privacy. The scheme authenticates pseudo IDs to verify the legitimacy of vehicles and employs threshold-based signature techniques. It suffers from high computation and communication overhead. These schemes showcase various approaches to identity-based authentication, each with strengths and limitations.

7.2.2 Identity-based authentication scheme without pairings

Due to the high computational cost associated with pairing operations, three times that of point multiplication operations [29] [30], validation plans without pairings have been proposed to streamline calculation time. Lo and Tsai [31] suggested an efficient conditional privacy-preservation scheme without pairings (ECPA). It utilizes an ECC-based IBS scheme and establishes secure vehicle-to-RSU (V2R) and RSU-to-Vehicle (R2V) communications. The scheme ensures message integrity and traceability. However, it lacks provisions for vehicle-to-vehicle (V2V) communication and is susceptible to external attacks, which are its main drawbacks. Cui et al. [32] introduce a secure privacy-preserving authentication scheme with cuckoo filter (SPACF). This scheme does not rely on tamper-proof devices and is primarily software-based. It utilizes cuckoo filters and binary search methods for efficient batch verification. The scheme provides message integrity, traceability, collision, and replying resistance. However, it is also prone to external attacks.

These schemes demonstrate efforts to develop identity-based authentication schemes without pairings to overcome the computational overhead. While they offer certain advantages, such as reducing computation time, they also have limitations related to inter-vehicle communication and vulnerability to external attacks. In [33], an efficient privacy-preserving authentication scheme (EPAS) for VANETs-based emergency communication is designed.

The scheme involves two trusted authorities (TAs), a trusted authority, and a disaster relief authority (DRA). It utilizes an ECC-based authentication scheme without pairings to establish secure V2D (vehicle to DRA) and V2V communications. However, its main drawback is the reliance on two trusted authorities. Additionally, it is insecure against external attacks.

Table 7.1 provides a comprehensive overview of various identity-based authentication schemes, comparing their features and communication patterns. It is evident from the table that most of the schemes lack security against external attacks, as they fail to achieve confidentiality. Furthermore, these schemes can only guarantee the ability to authenticate by relying on road side units (RSUs) during emergencies. To address these limitations, this study suggests a robust authentication scheme using ECC without pairings, specifically designed to counter external attacks.

7.3 Architecture and Requirements

7.3.1 System architecture

Figure 7.3 shows the architecture of the system. The figure shows that the vehicular environment has TAs, vehicles, and road side units (RSU). The whole organization will be isolated into numerous regions.

- **Trusted authority (TA):** Each region in the vehicular network is controlled by a TA. RSUs in the TA's region will be under its supervision. It additionally holds the task of vehicle pseudo ID generation during the registration process. Based on the two-layer vehicular model, TA is the upper layer, which controls the lower layer of RSUs and vehicles. TA is wholly trusted in the proposed scheme. It can never be compromised.
- **Road side unit (RSU):** They are integral to vehicular networks, providing services to vehicles and gathering traffic information. They broadcast information to vehicles within their range. RSUs are vulnerable to being exploited by attackers and are only partially trusted. They also play a crucial role in facilitating secure V2V (vehicle-to-vehicle) communications by broadcasting the public keys of vehicles to establish secure communication channels. Despite their importance, RSUs can be vulnerable to attacks, emphasizing the need for robust security measures to protect them and ensure the integrity of vehicular networks.
- **Vehicles:** The manufacturers will equip vehicles with OBU. It is imported with the public parameters of TA during the registration

Table 7.1 Comparison of existing ID-based schemes.

Scheme	Communication pattern	Cryptographic basis	Security features	Communication without RSU
[23]	V2R, R2V, & V2V	ID-based signature (IBS & IBOOS)	Conditional privacy-preservation and nonrepudiation	No
[24]	V2R, R2V, & V2V	ID-based signature (IBS & IBOOS)	Conditional privacy-preservation, nonrepudiation, and traceability	No
[25]	V2V, R2V, & V2R	ID-based signature (ECC)	Nonrepudiation, traceability, unlinkability, privacy-preservation, and batch verification	No
[26]	V2V, R2V, & V2R	ID-based signature (ECC)	Conditional unlinkability and signature aggregation	No
[27]	V2V, R2V, & V2R	Identity-based encryption (IBE)	Conditional privacy-preservation	No
[28]	V2V, R2V, & V2R	Identity-based threshold signature	Conditional privacy-preservation of confidentiality, message integrity, and nonrepudiation	No
[31]	V2R & R2V	ID-based signature (ECC)	Privacy-preservation, message integrity, traceability, and batch verification	No
[32]	R2V & V2R	ID-based signature (ECC)	Privacy-preservation, traceability, and message integrity	No
[33]	V2D & V2V	ID-based signature (ECC)	Conditional privacy-preservation, message integration, and batch verification	Yes

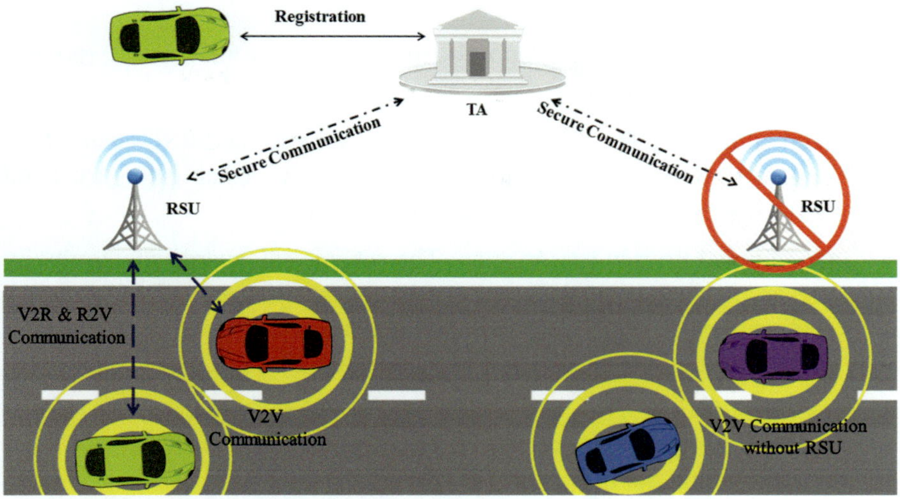

Figure 7.3 System architecture.

process. Vehicles will send messages regarding traffics, road condition, etc., to nearby RSUs. It also communicates with other vehicles.

RSUs and TAs are connected securely. Vehicles and TAs will communicate through a wireless connection. Inter-vehicle communication takes place within a dedicated short range.

7.3.2 Attack illustration

Two types of attacks are possible on the vehicular network. They are internal attacks and external attacks. In internal attacks, the attackers will be within the network and can access all parameters used for signature generation. They can perform the following malicious acts: alteration of transmitting messages, sending fake traffic-related messages, masquerading as a legitimate vehicle, etc. Efficient authentication schemes can identify these attackers. External attackers are outsiders, and they are challenging to identify. Even though they cannot access the parameters, they can observe and analyze the traffic throughout the network. External attackers will also combine to explore the entire vehicular network traffic.

Few internal and external attackers are placed manually in the network. They try to observe and alter traffic-related messages. Legitimate vehicles can detect and isolate the attacker with the help of RSUs. This chapter focuses on external attackers since they are challenging to identify.

7.3.3 Security requirements

The framework that is being proposed aims to meet several security requirements, which are as follows:

- Authentication: The scheme ensures that vehicles and RSUs can differentiate between attackers and legitimate vehicles, establishing trust and verifying the identity of communicating entities.
- Conditional privacy-preservation: The scheme safeguards anonymity by preserving their original IDs and preventing the disclosure of private information to other vehicles during communication. However, in the event of violations, the scheme allows tracing back the actual IDs of vehicles.
- Nonrepudiation: The system guarantees vehicles cannot reverse activities after infringing. They are held accountable for their behavior and actions within the network.
- Message integrity: The scheme guarantees that the messages received by vehicles are identical to those transmitted without any unauthorized alteration or tampering during transmission.
- Confidentiality: The scheme ensures that the messages transmitted by a vehicle can only be accessed and understood by that specific vehicle. Other vehicles within the network cannot eavesdrop on or intercept the transmitted messages.
- Unlinkability: The scheme protects the identity of vehicles by preventing the linkage of a series of messages to identify a particular vehicle. This enhances the anonymity within the network.

The proposed plan aims to create a safe and reliable communication environment for vehicles and RSUs in the vehicular network by meeting these security requirements.

7.4 Proposed Scheme

There are five steps in the suggested framework: setup, registration, V2R and R2V communication, V2V communication, and V2V communication without RSU. The notations used are given in Table 7.2.

7.4.1 Setup

During the system setup stage, the initialization process takes place. The trusted authorities (TAs) generate system parameters that will be preloaded in all RSUs. These parameters are as follows:

Table 7.2 Notations.

Notation	Description
P	Generator point
A	TA's private key
X	TA's public Key
IDX, ID_R	Original ID of vehicle "x" and RSU
U_x, U_R	Public key of vehicle "x" and RSU
u_x, u_R	Private key of vehicle "x" and RSU
$PSID_x$	Pseudo ID of vehicle "x"
H, H_1	One-way hash functions
Vt	Valid time period of pseudo ID
T	Current timestamp
c_i	Random number
M	Traffic-related message
Z_x	Signature of vehicle "x"

- Finite field: The finite field F_n is generated, where the field size of F_n is represented by the huge prime integer n.
- Elliptic curve: For F_n, an elliptical curve E is defined. The equation $y^2 = x^3 + ax + b \pmod{n}$ is used, $(a, b) \in F_n$ represents the curve's parameters. It is important to note that $4a^3 + 27b^2 \neq 0$ ensures the curve's non-singularity.
- Elliptic curve group: The elliptical curve group G is established, and a generator point P is selected on the curve E. This generator point P is distinct from the point at infinity (∞).
- TA's private and public keys: Each TA generates a private key $\alpha \in Z_r$, with prime number r. TA multiplies its private key by the generator point P to create its equivalent public key X, i.e., $X = \alpha P$.
- Hash functions: Two one-way hash functions are chosen: H: $\{0,1\}^* \rightarrow Z_r$ and H_1: $\{0,1\}^* \rightarrow Z_r$. These hash functions are utilized for various cryptographic operations in the scheme.
- The public parameters $\{P, X, H_1\}$ are then preloaded into all RSUs, ensuring they have the necessary information to carry out secure communications within the system. These parameters form the basis for cryptographic operations and authentication mechanisms in the proposed scheme.

7.4.2 Registration

During registration, pseudo IDs are generated for vehicles instead of using their original IDs. These pseudo IDs are derived from the vehicle's personal information, like the name of the driver, plate number of the vehicle, etc. Using cryptographic techniques, the scheme ensures the privacy and protection of sensitive details. The generation of pseudo IDs maintains confidentiality while enabling secure and authenticated communication within the vehicular network.

Step 1: When a vehicle, "x," enters a specific zone, it sends its original ID (ID_x) to TA to register.

Step 2: After receiving ID_x, TA will verify it, and then it will generate multiple pseudo IDs by using the vehicle x's original ID if it is valid as follows:

$$\mathrm{PSID}_{xi} = H\,(\alpha ID_x,\ vt,\ C_i)\,X, \tag{7.1}$$

where α is TA's private key, ID_x is the vehicle's unique ID, vt is the valid period of the pseudo ID, and C_i is the random number ($i = 1, 2, \ldots, n$). Thereby "n" number of pseudo IDs is generated for vehicle "x."

Step 3: TA will also assign the vehicle x's private key, u_x. It then sends generated "n" pseudo IDs, a private key of vehicle x, and TA's public parameters $\{\mathrm{PSID}_i, vt, U_X, G, X, H_1\}$ to vehicle "x" through a secure wired connection.

Step 4: Vehicle "x" computes its public key using its private key as follows:

$$U_x = u_x P. \tag{7.2}$$

This way, multiple vehicle pseudo IDs are generated during registration. These pseudo IDs are used only once during signature generation.

7.4.3 R2V and V2R communication

TA will assign IDs (ID_R) and private keys (u_R) to RSUs within its range. RSUs compute their public key using its private key, u_R, as follows:

$$U_R = u_R P. \tag{7.3}$$

RSUs broadcast their public key to vehicles within its range of the form $\{ID_R, U_R, T, SIG_R\}$, where ID_R is RSU's ID, U_R is its public key, T is the current timestamp, and SIG_R is the RSU's signature, which is computed as follows:

$$SIG_R = H_1(ID_R, U_R, T). \tag{7.4}$$

7.4.3.1 Signing

After receiving broadcast messages from RSU, vehicle "x" will sign the traffic-related message using the RSU's public key as follows:

$$R_x = u_x U_R \tag{7.5}$$

$$V_x = M_x + R_x. \tag{7.6}$$

Vehicle "x" automatically creates its signature utilizing its one-time pseudo ID.

$$Z_x = H_1 (U_x, \mathrm{PSID}_{xi}, M_x, T) U_R + R_x. \tag{7.7}$$

Vehicle "x" sends {PSID_{xi}, V_x, Z_x, U_x, vt, T} to a nearby RSU. This is known as V2R communication.

7.4.3.2 Verification

RSU checks the current timestamp value and vt's validity. The RSU then uses the subsequent equations to determine the authenticity of the vehicle.

$$Y_R = u_R U_x \tag{7.8}$$

$$M = V_x - Y_R \tag{7.9}$$

$$Z_x = H_1 (U_x, \mathrm{PSID}_{xi}, M_x, T) U_R + Y_R. \tag{7.10}$$

The message is accepted by RSU and sent to all vehicles within its range if eqn (7.10) is true. Vehicle "x" will receive an acknowledgment message of the form {ID_R, ack, U_i, SIG_R} from RSU along with the public keys of all other vehicles in its range because it is the legitimate vehicle. After verifying RSU's signature, vehicle x will accept the list of public keys and will use it to establish secure V2V communications.

7.4.4 V2V communication

Vehicles will communicate with other vehicles in the region. It has to be carried out securely. Inter-vehicle communication is established with the help of R2V communication. Vehicle "x" will communicate with vehicle "y" using the following steps.

7.4.4.1 Signing

Vehicle "x" will first sign the traffic-related messages using the vehicle y's public key. It acquires the vehicle's public key from RSU by R2V communication. Message signing is carried out as follows:

$$R_x = u_x U_y \tag{7.11}$$

$$V_x = M_x + R_x. \tag{7.12}$$

Vehicle "x" generates its signature using its one-time pseudo ID.

$$Z_x = H_1 (U_x, PSID_{xi}, M_x, T) U_y + R_x. \tag{7.13}$$

Vehicle "x" sends {PSID_{xi}, V_x, Z_x, vt, T} to vehicle "y."

7.4.4.2 Verification

Vehicle "y" checks the present timestamp value and vt's validity. Then vehicle y checks the legitimacy of the vehicle x as follows:

$$Y_y = u_y U_x \tag{7.14}$$

$$M = V_x - Y_y \tag{7.15}$$

$$Z_x = H_1 (U_x, \mathrm{PSID}_{xi}, M_x, T) U_y + Y_y. \tag{7.16}$$

Vehicle "y" receives the message from vehicle "x" if eqn (7.16) holds.

7.4.5 V2V Communication without RSUs

During natural disasters, RSUs in that region will be damaged entirely; there is also a chance for an RSU to get damaged due to specific other reasons. In such cases, the vehicles should communicate with one another without RSUs. This type of communication during emergencies without RSUs is known as emergency communications. Messages transmitted during emergency communication regarding damaged road conditions, accidents, traffic, etc., are crucial; the network will be prone to various attacks. Hence, they should be transmitted securely. Emergency communications are carried out securely by following the given steps.

Step 1: Vehicle "x" will await the RSU's broadcast message. Since it does not receive broadcast messages, vehicle x will determine the RSU is damaged. It then computes its signature as

$$Z_x = H_1 (\mathrm{PSID}_{xi}, U_x, T). \tag{7.17}$$

Step 2: After computation, vehicle "x" will broadcast a request message with its public key $\{PSID_{xi}, Z_x, U_x, T\}$.

Step 3: Vehicle y will verify the vehicle x's signature, sign the traffic-related message with vehicle x's public key, and then deliver it to vehicle x.

Step 4: Vehicle "y" sends messages to vehicle "x," which verifies the message's signature and responds if it is legitimate. The signing and verification process is the same as in Section 7.4.4.

7.5 Security Analysis

7.5.1 Authentication

In the proposed scheme, vehicles transmit their signatures to other vehicles and RSUs. For instance, vehicle x sends its signature Z_x, computed using various parameters, including its private key and other relevant information. An attacker must obtain the private key value obtained by using the vehicle's public key through the elliptic curve discrete logarithm problem to forge the signature. This ensures the authenticity of communication in V2R, R2V, and V2V scenarios, as an attacker's attempt to extract the confidential key from the public key is computationally impossible. In this manner, the proposed framework gives authentication in these types of communications within the vehicular network.

7.5.2 Conditional privacy-preservation

TA generates vehicle pseudo IDs using a hash function, the vehicle's original ID, and the private key. These pseudo IDs protect the privacy of the vehicle's original ID from other entities. The TA stores the pseudo IDs in its memory for a specific validity period. If any violations occur, the original ID can be recovered using it. This approach achieves conditional privacy-preservation by revealing the original ID only when necessary for addressing violations or maintaining security. The scheme ensures that the vehicle's identity remains confidential while allowing the TA to identify vehicles when needed.

7.5.3 Nonrepudiation

The messages will be encrypted using sender's private and receiver's public key. This ensures the confidentiality and authenticity of the messages. The signature generation process uses the vehicle's pseudo ID and public key, reinforcing the nonrepudiation aspect. Once a message is signed and sent, the vehicle cannot deny its authenticity or claim that it did not send the message. This strengthens the system's overall security by ensuring the integrity and accountability of the communicated information.

7.5.4 Message integrity and confidentiality

The proposed scheme encrypts messages using the receiver's public key before transmission. This ensures that only the intended recipient – who also possesses the associated private key – can access and decrypt the message. As a result, attackers cannot modify the transmitted messages since they lack the necessary encryption keys. This guarantees message integrity, as any unauthorized modifications would render the decryption process unsuccessful. The encryption process also provides confidentiality by safeguarding the information from unauthorized access. Thus, the proposed scheme achieves both message integrity and confidentiality.

7.5.5 User unlinkability

One-time pseudo IDs and current timestamp values generate signatures. Attackers cannot link two or more signatures to a generated vehicle because pseudo IDs and the current timestamp value will differ for each signature. A vehicle will not generate the same signature. Hence, vehicles cannot be linked by their signatures, and the linkability [34], [35] problem does not abide.

7.5.6 Resistance to external attacks

The proposed scheme ensures confidentiality. Hence, external attackers, generally outsiders, cannot decrypt the messages transmitted from one vehicle to another. It cannot analyze the traffic in the network as well. Attacks based on external attacks like eavesdropping and passive man-in-middle attacks are also impossible.

7.6 Performance Analysis

In this section, an examination is made between the proposed framework and other frameworks concerning the calculation and correspondence above. Specifically, the focus is on comparing the proposed scheme with ID-based authentication schemes that utilize pairings. These schemes involve expensive pairing operations, resulting in high computation and communication

Table 7.3 Execution time of cryptographic operations.

	T_M	T_{MP}	T_P
Execution time (ms)	0.03	1.50	8.12

overhead. It is also contrasted with current techniques that do not rely on pairings. These schemes offer potential advantages in terms of reduced computational and communication requirements.

7.6.1 Simulation scenario

The vehicular environment is created with 125 mobile nodes, 3 RSUs, and 1 TA. 5000 m × 5000 m operating space is used as a simulation area in which vehicles move in the created environment. Vehicles are moved in the vehicular environment within 30 seconds. Communication occurs by exchanging packets between the sender and the recipient.

7.6.2 Computation overhead

The suggested scheme's computation time is compared to recent ID-based systems [31], [32], [33]. An Intel Pentium 4 computer with 1 GB of RAM and a 3.0 GHz processor is used to conduct the evaluation. The execution times of different cryptographic procedures within the framework are shown in Table 7.3. Specifically, the pairing operation's calculation time is represented by T_P, the scalar multiplication operation's calculation time is represented by T_M, and T_{MP} denotes the scalar multiplication point operation's computation time.

The results show that T_M takes 0.03 ms, T_{MP} takes 1.50 ms, and T_P takes 8.12 ms. These figures indicate the computational costs of each operation in the proposed scheme. Table 7.4 compares the computation times for different schemes to analyze the computation costs further. Additionally, T_H indicates the amount of time needed for a one-way hash function (0.0001 ms), and the one-point multiplication operation's calculation time is represented by T_{PA} (0.0018ms).

As shown in Table 7.4, signature generation has one T_H, two T_{MP}, and two T_{PA}. Hence, the computational time to generate a signature is 3.0037 ms, whereas it is 1.5301, 1.5001, 1.5019, and 3.0019 ms in ECPA [31], SCPF [32], EPAS: Scheme 1 [33], and EPAS: Scheme 2 [33], respectively (refer Table 7.3). It is clear from Table 7.4 that the signature generation in the proposed scheme incurs high computational costs. This is due to the encryption

Table 7.4 Computation cost.

	Signature generation	**Signature verification**
ECPA [31]	$T_{MP} + T_M + T_H$	$3T_{MP} + 3T_{PA} + 2T_H$
SPACF [32]	$T_{MP} + T_H$	$2T_{MP} + T_{PA} + T_H$
EPAS: Scheme 1 [33]	$T_{MP} + T_{PA} + T_H$	$3T_{MP} + 2T_H$
EPAS: Scheme 2 [33]	$2T_{MP} + T_{PA} + T_H$	$2T_{MP} + T_H$
Proposed scheme	$2T_{MP} + 2T_{PA} + T_H$	$2T_{MP} + T_{PA} + T_H$

of the traffic-related message before sending it to its neighbors. Since the suggested system achieves confidentiality while protecting against external attacks, the high computational cost of signature generation is highly acceptable.

Signature verification has two T_{MP}, one T_H, and one T_{PA}. Hence, the computational time taken for one signature verification is 3.0019 ms, whereas it is 4.5056, 3.0019, 4.5002, and 3.0001 ms in ECPA [31], SCPF [32], EPAS: Scheme 1 [33], and EPAS: Scheme 2 [33], respectively (refer Table 7.3). The proposed scheme has less computational cost than ECPA [14] EPAS: Scheme 1 [33] and the exact computational cost as that of SCPF [32] EPAS: Scheme 2 [33]. It is understandable from the table that even though the proposed scheme has decryption operations in addition to the verification process, compared to other systems, it is more computationally cheap for signature verification.

The computational delay is calculated by using the following expression:

$$CD = T_{\text{Sender Sign}} + T_{\text{Receiver Verify}}. \tag{7.18}$$

The proposed framework's computational delay is 6 ms, whereas it is 6, 4.5, 6, 6, and 6 ms in ECPA [31], SCPF [32], EPAS: Scheme 1 [33], and EPAS: Scheme 2 [33], respectively. The suggested scheme's computational delay is similar to previous techniques except for the SCPF [32] scheme.

As shown in Figure 7.4, the computation time of pseudo IDs is compared with the ECPA [31] scheme since TA generates multiple pseudo IDs for each vehicle in both schemes. It is clear from the graph that the computation time (average) is 50% less than the ECPA [31] scheme.

Table 7.5 contrasts the calculation cost of the proposed scheme's signature verification method with that of other ID-based schemes. Compared to ECPA [31], the suggested framework's percentage improvement is approximately 33.37%. However, the proposed scheme shows no improvements compared to SCAPF [32] and EPAS: Scheme 2 [33], as they have the exact computational cost for signature verification.

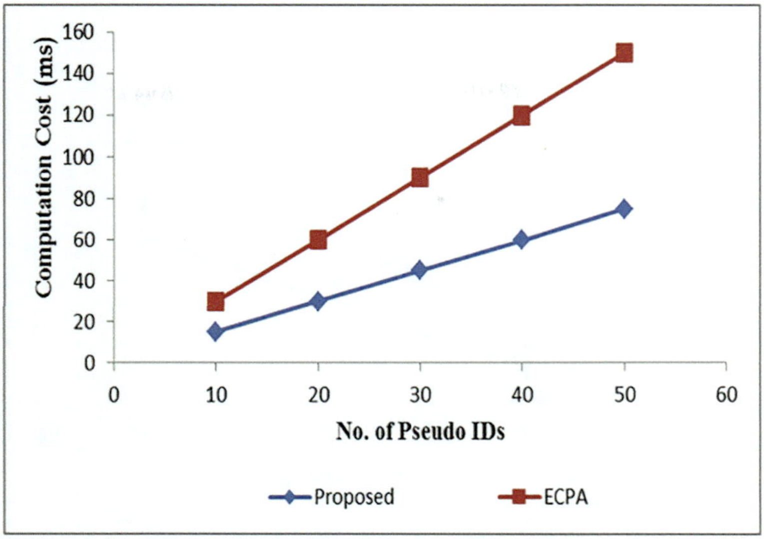

Figure 7.4 Computation time for pseudo IDs.

Table 7.5 Comparison of computation cost.

Scheme	Signature verification
ECPA [14]	33.37%
SCAPF [8]	0%
EPAS: Scheme 1 [22]	33.29%
EPAS: Scheme 2 [22]	0%

As shown in Table 7.5, the proposed scheme exhibits a 33% improvement over EPAS: Scheme 1 [33] and ECPA [31] regarding signature verification. This indicates that, in comparison to previous methods, the suggested scheme provides more effective signature verification. Although the proposed method requires more processing time to generate a signature, it is considered acceptable since it provides security against external attacks. Moreover, the proposed scheme demonstrates lower computation time for signature verification and computation delay. The pseudo ID computation cost is also lower in the proposed scheme.

7.6.3 Communication overhead

This section compares the proposed scheme's communication costs to competing schemes. The proposed framework has less transmission overhead than the other schemes, as shown in Figure 7.5. Specifically, the proposed scheme

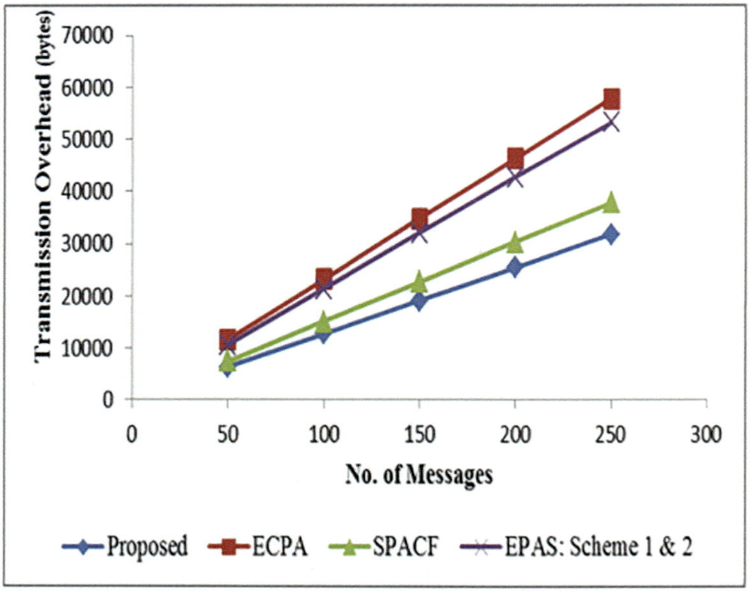

Figure 7.5 Transmission overhead.

exhibits a 15% reduction in transmission overhead compared to the SCPF [32], resulting in 24 bytes less transmitted data. Furthermore, Figure 7.3 shows that the proposed IBS scheme has a 45% reduction in transmission overhead compared to the ECAP [31], resulting in 104 bytes less transmitted data. Additionally, the proposed scheme demonstrates a 40% reduction in transmission overhead compared to EPAS: Schemes 1 and 2, resulting in 86 bytes less transmitted data.

The section also discusses communication overhead in terms of the parameter "delay," which represents the contrast between the received time and the sent season of a packet. This parameter determines the transmission time of packets.

It is computed as follows:

$$\text{Delay} = \text{Packet received time} - \text{Packet sent time}. \quad (7.19)$$

This formula can calculate how long a packet travels from its source to its destination. It helps to determine the communication overhead in the network. The delay in R2V and V2R communication is compared to the ECPA scheme. In Figure 7.6, the delay is less than the ECPA scheme. R2V is 16% less than ECPA, whereas V2R delay is 1.3% less than ECPA scheme. Therefore, as demonstrated, the suggested framework has reduced delay.

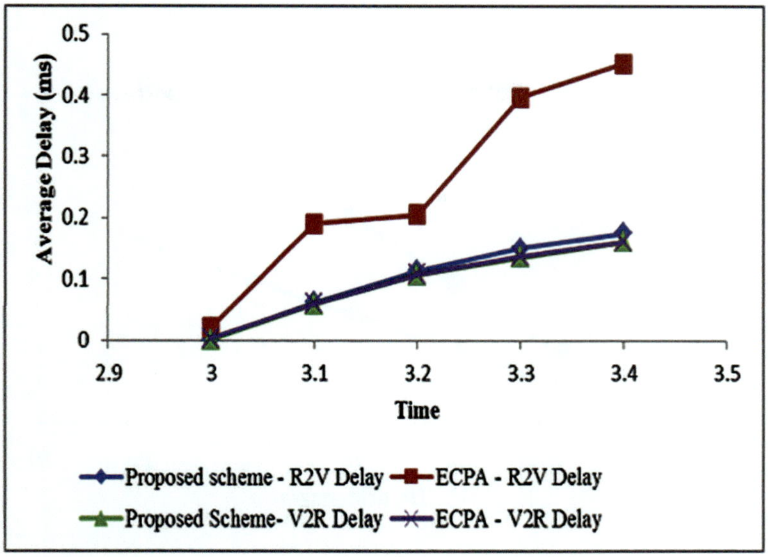

Figure 7.6 R2V and V2R delay.

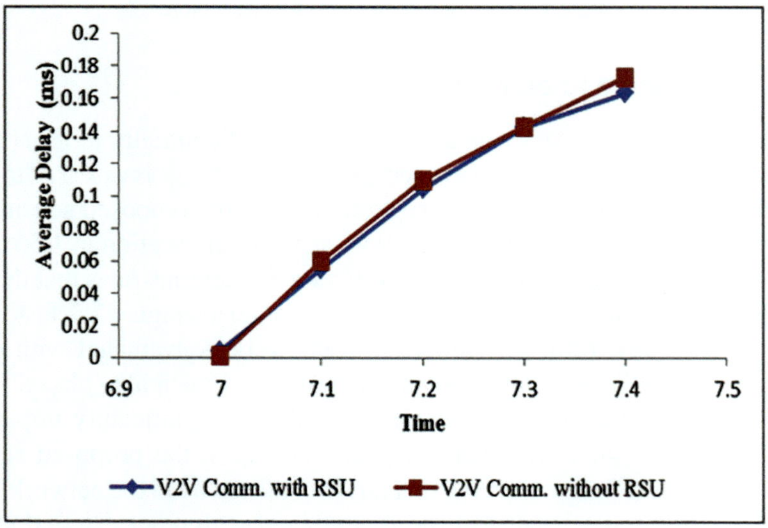

Figure 7.7 V2V delay.

V2V communication with and without RSU is contrasted in Figure 7.7. The delay in V2V communication without RSU is 3.4% less than V2V communication with the RSU. Even in the absence of RSU, the delay is less. Hence, as analyzed, the communication overhead is less.

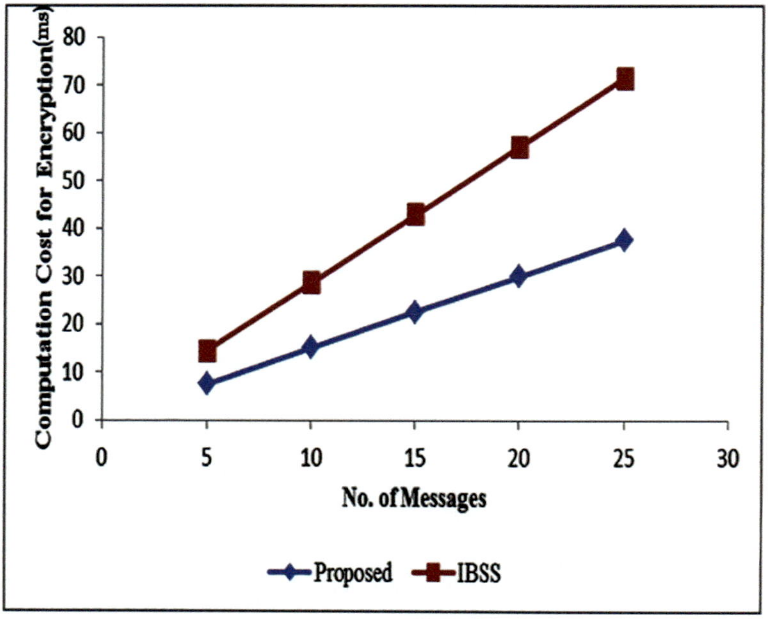

Figure 7.8 Computation cost for encryption.

7.6.4 Resistance to eternal attacks

The proposed framework focuses on achieving confidentiality to resist external attackers. The time required to encrypt a single message is assessed to compare the computational cost of encryption between the proposed scheme and the scheme [28]. The proposed framework achieves encryption in 1.5018 ms, while the scheme [28] requires 2.86 ms. Figure 7.8 demonstrates that the suggested scheme's encryption computation time is, on average, 47% faster than that of the scheme [28]. The proposed framework is evaluated with external attacks in NS2 simulation, where attackers are manually placed in the network. External attackers are passive and do not significantly impact the scheme's performance. By encrypting the messages, the proposed scheme prevents attackers from observing or analyzing the traffic in the network. They cannot perform traffic analysis even if combined with other external attackers. Accordingly, the proposed framework is secure against outside assaults.

7.7 Conclusion

This chapter proposes a secure identity-based authentication scheme without pairings against external attacks. The scheme leverages ECC to provide efficient authentication while ensuring confidentiality to achieve security against external attacks. The security investigation shows that it fulfills all fundamental security necessities. One of the notable advantages of the proposed scheme is its low computation overhead for signature verification compared to other schemes. It also exhibits a low computational delay, enhancing its efficiency. Although the scheme incurs a higher computational cost for signature generation, this trade-off is acceptable due to its resilience against external attacks. Additionally, the time taken to encrypt a message is equivalent to ECC's point multiplication, contributing to the scheme's low computation cost for ensuring confidentiality and providing security against external attacks.

As a future direction, the proposed scheme could explore the implementation of batch verification, which allows for the simultaneous verification of multiple messages from different vehicles. This enhancement would further increase the scheme's scalability and efficiency.

References

[1] Leelavathy, S., Nithya, M., Dhaya, R., Muthuselvan, S., Kanthavel R., and Rajakumari, K. "Effective Traffic Model for Intelligent Traffic Monitoring Enabled Deep RNN Algorithm for Autonomous Vehicles Surveillance Systems," 2023 4th International Conference on Smart Electronics and Communication (ICOSEC), Trichy, India, 2023, pp. 1191-1200

[2] Viadero-Monasterio, F., Nguyen, A. -T., Lauber, J., Boada M.J.L., and Boada, B.L. "Event-Triggered Robust Path Tracking Control Considering Roll Stability Under Network-Induced Delays for Autonomous Vehicles," in IEEE Transactions on Intelligent Transportation Systems

[3] Zhao, H., Yang, H., Xia Y., and Zuo, Z. "Nonlinear Terminal-Free MPC on Multitype Bend Tracking With Discontinuous Reference Paths for Autonomous Vehicles," in IEEE Transactions on Industrial Electronics, vol. 71, no. 1, pp. 810-821, Jan. 2024

[4] https://subscription.packtpub.com/book/iot-and-hardware/9781838649326/10/ch10lvl1sec93/components-of-a-typical-self-driving-car

[5] Azimi, R., Bhatia, G., Rajkumar, R., and Mudalige, P. "Vehicular networks for collision avoidance at intersections," Proc. SAE World Congr., Detroit,MI, USA, Apr. 2011, pp. 1–11.

[6] Zeadally, S. et al., "Vehicular Ad Hoc Networks (VANETS): Status, Results, and Challenges," Telecomm. Systems, vol. 50, no. 4, pp. 217–241, 2012.

[7] Luan, et al., "Autonomous Vehicular Networks: Perspective and Open Issues", Networking and Internet Architecture.
[8] Misener, J.A. "Vehicle-infrastructure integration (VII) and safety: Rubber and radio meets the road in California," Intellimotion, vol. 11, no. 2, pp. 1–3, 2005.
[9] Lee, U. et al., "Mobeyes: Smart mobs for urban monitoring with a vehicular sensor network," IEEE Wireless Commun., vol. 13, no. 5, pp. 52–57, Oct. 2006.
[10] Armknecht, F. et al., "Cross-Layer Privacy Enhancement and Nonrepudiation in Vehicular Communication," Proc. ITG-GI Conf.Comm. in Distributed Systems (KiVS), pp. 1-12, 2007.
[11] Fuentes, J.M.D. et al.,"Overview of Security Issues in Vehicular Ad-Hoc Networks," Handbook of Research on Mobility and Computing, pp. 894–911, IGI Global Snippet, 2011.
[12] Raya, M. et al., "Securing Vehicular Ad hoc Networks," J. Computer Security, vol. 15, no. 1, pp. 39–68, 2007.
[13] Shamir, A. "Identity-based cryptosystems and signature schemes," in Advances in Cryptology. New York, NY, USA: Springer-Verlag, 1984, pp. 47–53.
[14] Yeh, L.Y., and Lin, Y.C. "A proxy-based authentication and billing scheme with incentive-aware multihop forwarding for vehicular networks," IEEE Trans. Intell. Transp. Syst., vol. 15, no. 4, pp. 1607–1621, Aug. 2014.
[15] Jenefa, J., and Mary Anita, E.A. Secure Authentication Schemes for Vehicular Adhoc Networks: A Survey. Wireless Pers Commun 123, 31–68 (2022). https://doi.org/10.1007/s11277-021-09118-3
[16] Jenefa, J., and Anita, E.A.M. Identity-based message authentication scheme using proxy vehicles for vehicular ad hoc networks. Wireless Netw 27, 3093–3108 (2021).
[17] Mary Anita, E.A., Lakshmi, S., and Jenefa, J. A self-cooperative trust scheme against black hole attacks in vehicular ad hoc networks, Vol. 21 No. 1, pp. 59–65.
[18] Lakshmi. S, Mary Anita E.A., and Jenefa, J. Detection and Prevention of Black Hole Attacks in Vehicular Ad Hoc Networks, Volume-8, Issue-7, May 2019.
[19] Jenefa, J., and Mary Anita, E.A. An Enhanced Secure Authentication Scheme for Vehicular Ad Hoc Networks Without Pairings. Wireless Pers Commun 106, 535–554 (2019).
[20] Jenefa, J., and Mary Anita, E.A. Secure Vehicular Communication Using ID Based Signature Scheme. Wireless Personal Communication (2017).

[21] Anita E.A.M., and Jenefa, J. "A survey on authentication schemes of VANETs," 2016 International Conference on Information Communication and Embedded Systems (ICICES), Chennai, India, 2016, pp. 1–7.

[22] Jenefa, J., Sajini, S., and Anita, E.A.M. (2023). Secure IBS Scheme for Vehicular Ad Hoc Networks. In: Reddy, V.S., Prasad, V.K., Wang, J., Reddy, K.T.V. (eds) Soft Computing and Signal Processing. ICSCSP 2022. Smart Innovation, Systems and Technologies, vol 313. Springer, Singapore.

[23] Jie Li, Huang Lu, and Mohsen Guizani. "ACPN: A Novel Authentication Framework with Conditional Privacy-preservation and Nonrepudiation for VANETs", IEEE Transactions on Parallel and Distributed Systems, vol. 26, no. 4, pp. 938–948, April 2015.

[24] Cong Sun, Jioa Liu, Xinpeng Xu, and Jianfeng Ma. "MADAR: A Privacy-Preserving Mutual Authentication Resting DoS Attacks in VANETS", IEEE Access, vol. 5, pp. 24012–24022, November 2017.

[25] Shiang-Feng Tzeng,Shi-Jinn Horng,Tianrui Li, Xian Wang, Po-Hsian Huang, and Muhmmad Khurram Khan. "Enhancing Security and Privacy for Identity-based Batch Verification Scheme in VANET", IEEE Trans. Veh. Technol.

[26] Lei Zhang, Qianhong Wu, Josep Domingo-Ferrer, Bo Qin, and Chuanyan Hu. "Distributed Aggregate Privacy-Preserving Authentication in VANETs", IEEE Transactions on Intelligent Transportation Systems.

[27] Dijiang Huang, Satyajayant Misra, Mayank Verma, and Guoliang Xue. "PACP: An Efficient Pseudonymous Authentication-Based Conditional Privacy Protocol for VANETs", IEEE Transactions on Intelligent Transportation Systems, vol. 12, no. 3, September 2011.

[28] Jinyuan Sun, Chi Zhang, Yanchao Zhang, and Yuguang Fang. "An Identity-Based Security System for User Privacy in Vehicular Ad Hoc Networks", IEEE Transactions on Parallel and Distributed Systems, vol. 21, no. 9, September 2010.

[29] Barreto, P.S.L.M., Kim, H.Y., Lynn, B., and Scott, M. "Efficient algorithms for pairing-based cryptosystems," in Proc. Crypto, 2002, pp. 354–368.

[30] Tsai, J.L., and Lo, N.W. "A privacy-aware authentication scheme for distributed mobile cloud computing services," IEEE Syst. J., vol. 9, no. 3, pp. 805–815, Sep. 2015.

[31] Nai-Wei Lo, and Jia-Lun Tsai. "ECPA: An Efficient Conditional Privacy-Preservation Scheme for Vehicular Sensor Networks Without Pairings", IEEE Transactions on Intelligent Transportation Systems, vol. 17, no. 5, pp. 1319–1328, May 2016.

[32] Jie Cui, Jing Zhang, Hong Zhong, and Yan Xu. "SPACF: A Secure Privacy-preserving Authentication Scheme for VANET with Cuckoo Filter", IEEE Trans. Veh. Technol., vol. 14, no. 8, August 2015.
[33] Xuedan Jia, Xiaopeng Yuan, Lixia Meng, and Liangmin Wang. "EPAS: Efficient Privacy-preserving Authentication Scheme for VANETs-based Emergency Communication", Journal of Software,vol. 8, no. 8, August 2013.
[34] Tsai, J.L., Lo, N.W., and Wu, T.C. "Novel anonymous authentication scheme using smart cards," IEEE Trans. Ind. Informat., vol. 9, no. 4, pp. 2004–2013, Nov. 2013.
[35] Yan, G., Wen, D., Olariu, S., and Weigle, M.C. "Security challenges in vehicular cloud computing," IEEE Trans. Intell. Transp. Syst., vol. 14, no. 1, pp. 284–294, Mar. 2013.

8

Internet of Drones Security and Privacy

S. Sophia, Getzi Jeba Leelipushpam Paulraj, and Immanuel Johnraja Jebadurai

Department of Computer Science and Engineering, Karunya Institute of Technology and Sciences, India

Email: sophia@karunya.edu; getz23@gmail.com; immanueljohnraja@gmail.com

Abstract

The usage of unmanned aerial vehicles has increased in the recent past in various applications due to their ability to transport objects, take images, and feed live videos. Incorporated with the Internet of Things, the unmanned aerial vehicle has been renamed the Internet of Drones. Applications of the Internet of Drones include traffic surveillance, defense, agriculture crops monitoring, cinematography, underwater purposes, healthcare, logistics, commercial aids, GIS (geographic information system) mapping, environmental management, smart cities, disaster management, search and rescue operations, and various military applications. The decentralized network and management architecture of drones are connected and controlled to form the Internet of Drones. In the Internet of Drone environment, a network made up of numerous drones transmits and receives data from one another. A lack of security makes it possible for an attacker to exploit this vulnerability and manipulate or abuse sensor inputs or functions, or deactivate them to launch cyber-attacks and produce unpleasant failsafe measures. The possibility and frequency of these attacks are very high, which affects the privacy, confidentiality, availability, integrity, and trust of drone data. Thus, the security and privacy threats have to be addressed and it is necessary to take preventive and protective countermeasures. This chapter surveys the possible cyber-attacks and the security against such attacks. Also, it comes up with new technologies and techniques to overcome cyber-attacks on the Internet of Drones. Finally, it guides future research to help in the identification of the most recent Internet of Drone research opportunities.

8.1 Introduction

The Internet of Things has become an emerging cutting-edge environment including the areas of civilian, commercial, and military purposes [1]. IoT (Internet of Things) devices have a lot of benefits for improving services in many different contexts. The number of networked devices is expected to reach incredible heights in the coming years. The academic and professional communities have recently shown a great deal of interest in unmanned aerial vehicles. The usage of drones has become increasingly popular in recent years due to advancements in technology and their growing range of applications. They are used extensively in the field of industries, including agriculture, construction, entertainment, media, surveillance, and package delivery [2]. Drones are also used in emergency response and disaster relief efforts, search and rescue missions, and wildlife monitoring. Drones can record high-quality videos and photographs from a variety of perspectives since they are equipped with a wide range of sensors and cameras. They may be used to inspect the infrastructure, check the health of the crops, and even deliver packages or goods to distant places. Drones, which are flying machines with no onboard human pilot, have garnered a lot of public interest. To save money on helicopters and pilots, news organizations would like to use drones to record traffic and other important occurrences. Retailers are increasingly opting for drones to transport goods as it significantly cuts down on the cost of using vehicles and drivers. Additionally, drones are essential for individuals to engage in various adventurous and sports activities [3]. However, the usage of drones has also raised concerns about privacy, security, and safety. Regulations and guidelines have been put in place to address these concerns and ensure that drones are used safely and responsibly.

A network of connected unmanned aerial vehicles (UAVs) working together to perform a certain mission is known as the Internet of Drones, or IoD. IoD is created by replacing "Things" with "Drones" while maintaining a great deal of the same characteristics as IoT. It is anticipated that IoD will play a significant role in the advancement of UAVs. IoD is a "layered network control architecture" that helps UAVs communicate with one another. In the IoD environment, a network made up of numerous drones transmits and receives data from one another. IoD provides the capability of using IP (internet protocol) addresses to operate remotely or via the Internet [4]. The IoD enables drones to communicate with each other, share information, and coordinate their actions in real time. It also allows drones to interact with other connected devices, such as sensors, cameras, and servers, to perform tasks more efficiently. For example, in agriculture, drones equipped with sensors can collect data on crops, such as soil moisture, temperature, and plant health, and transmit that information to a centralized system for analysis. The system can then use that information to optimize crop management, reduce waste, and increase yield. The IoD has the potential to transform businesses by offering more precise, effective, and resourceful solutions. It also brings up issues with security, data privacy, and regulation. To ensure the safe and ethical usage of drones in the IoD, it is crucial to develop rules and guidelines for their use.

The Internet of Things (IoT) has gained momentum in various areas, such as leveraging secure edge computing through IoT, integrating IoT with 5G/6G networks, enhancing smart transportation systems with IoT, exploring different educational aspects of IoT and edge computing, employing blockchain to improve IoT network security, utilizing artificial intelligence for data analysis in IoT systems, implementing IoT-based monitoring and data recovery for analytical purposes, fortifying IoT systems against cyber–physical threats, and implementing blockchain solutions [5].

8.2 Applications of Internet of Drones

The Internet of Drones has many potential applications and is summarized as follows. Some of the applications of the IoD are shown in Figure 8.1 [6].

Delivery services: Drones can deliver goods, medicine, and food to people in far-off places or crowded places. The IoD can aid communication and coordination across numerous drones in the distribution organization to ensure speedy delivery.

Agriculture: Drones equipped with sensors and cameras are used to monitor crops, soil (Figure 8.2), and weather conditions to help ranchers make informed crop management decisions. Through the Internet of Things, drones may share information and control their development, enabling them to cover larger areas and provide more in-depth information.

Search and rescue: Using an IoD system, multiple drones can collaborate over large areas and exchange real-time data to increase the likelihood of quickly locating missing persons during search and rescue missions.

Inspection and maintenance: Drones have the capability to inspect and maintain buildings, bridges, and electricity lines. They are capable of swiftly

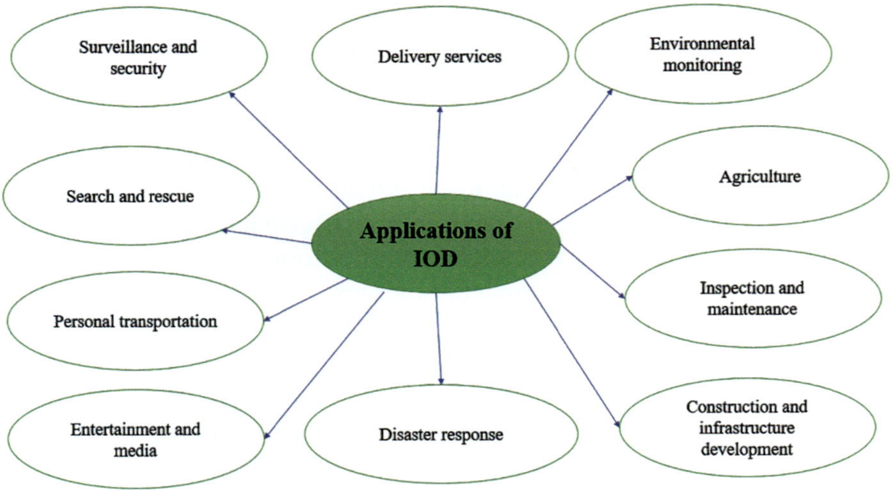

Figure 8.1 Applications of IoD.

and effectively examining vast areas by sharing data and coordinating their movements with the help of the Internet of Things (IoT). As a result, the need for inspections that require human intervention is eliminated.

Surveillance and security: Drones are useful for various purposes in the field of security and surveillance (Figure 8.2). They can be deployed to observe crowds at public events, monitor borders, and safeguard wildlife. Drones can efficiently cover large areas, communicate information, and coordinate movements to respond promptly to any catastrophe or threats.

Environmental monitoring: Drones can serve as a useful tool for observing weather patterns and collecting information on wildlife populations, alterations in land use, and the quality of air and water. With the help of the Internet of Drones (IoD), drones can exchange data and streamline their development to efficiently cover greater areas, offering valuable insights into everyday issues and aiding in strategic decision-making.

Disaster response: Drones are increasingly being used in disaster response (Figure 8.2) efforts to evaluate the extent of damage, identify survivors, and transport essential supplies. The Internet of Drones (IoD) can facilitate different drones to collaborate with each other, enabling them to cover larger areas and provide continuous information. This can help improve the effectiveness of disaster response efforts and enable a more timely and efficient response to disaster-stricken areas.

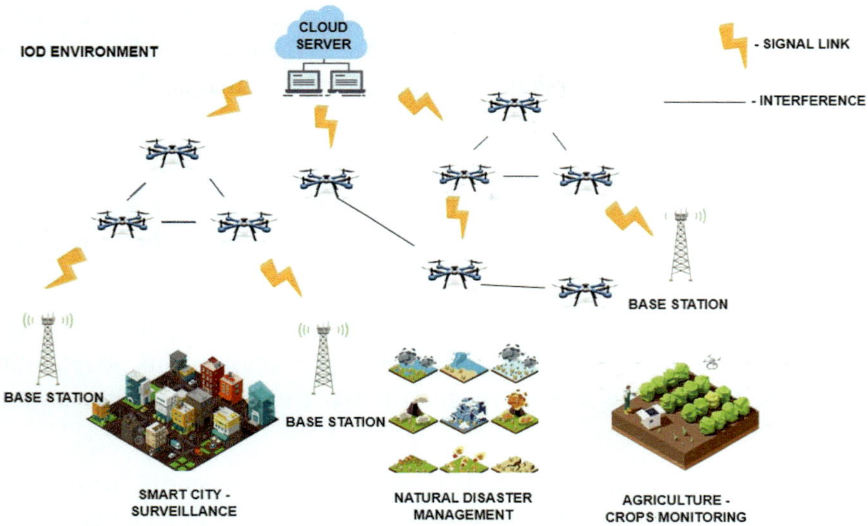

Figure 8.2 An ideal illustration of IoD in various fields.

Construction and infrastructure development: Drones can be utilized in construction projects to map sites, monitor progress, and inspect for quality and safety. Through the use of the Internet of Things (IoT), drones can communicate data and coordinate their movements, enhancing accuracy and efficiency in building and infrastructure development.

Entertainment and media: Drones can capture aerial video for use in various media and sports. Multiple drones can work with IoT to provide dynamic footage from different angles and perspectives.

Personal transportation: Drones can be used for personal transportation and to ensure safe and efficient delivery through IoT-connected communication.

8.2.1 Key characteristics of IoD

The Internet of Drones (IoD) has several key characteristics that set it apart from other applications of drone technology. These characteristics include:

Connectivity: A network of interconnected drones with the ability to exchange data and connect is essential to the Internet of Things. This network enables drones to work together to achieve common goals and continuously provide data.

Autonomy: Drones used by the IoD are autonomous, meaning they can carry out tasks without assistance from people. Due to their independence, drones can do difficult tasks and operate in hazardous environments where a human presence could be unrealistic or dangerous.

Data exchange: To enable coordination and cooperation, the IoD relies on drones to exchange information. This information exchange may include specifics like distance, speed, altitude, battery life, and other significant measurements.

Scalability: The IoD is designed to be flexible, which implies that it can accommodate a huge variety of drones and applications. This adaptability enables the IoD to assist a wide range of purpose cases and adapt over time to changing needs.

Security: The Internet of Things (IoT) requires strong security measures to prevent cyber-attacks and ensure the privacy and integrity of data transmitted between drones. This security includes encryption, authentication, and other steps to prevent unauthorized network access.

Flexibility: The IoD is designed to be flexible, which implies that it can change as conditions and requirements do. Drones can respond to shifting environmental conditions, unexpected obstacles, and other challenges that may arise during operation because of their versatility.

Interoperability: To enable effective coordination and cooperation, the IoD requires compatibility across various types of drones and correspondence systems. This interoperability ensures that drones can communicate with one another and exchange information, regardless of the manufacturer or messaging protocol used.

8.2.2 Taxonomy of Internet of Drones

The Internet of Drones (IoD) can be classified into different taxonomies [7] based on various factors are shown in Figure 8.3. Here are a few possible taxonomies of the IoD:

Based on communication:
Connected drones: These are drones that can communicate with other devices, like smartphones or computers, and are connected to the Internet.

Autonomous drones: These are autonomous drones that can make decisions based on their programming and operate without human intervention.

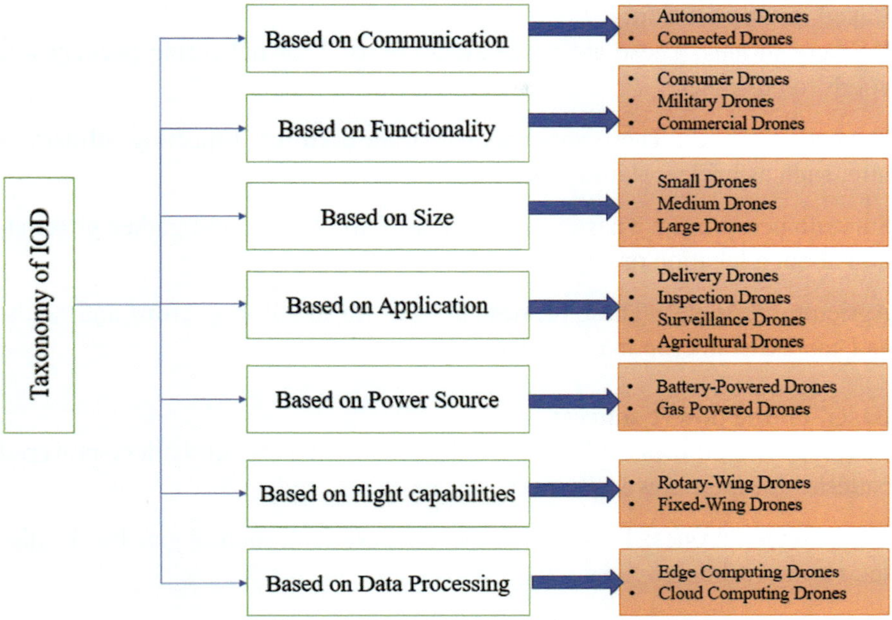

Figure 8.3 Taxonomy of Internet of Drones.

Based on functionality:
Consumer drones: These are the drones used by individuals for personal use, such as for aerial photography or racing.

Commercial drones: These are the drones utilized by organizations for different purposes, for example, delivering packages or inspecting infrastructure.

Military drones: These are the drones used by the military for surveillance, reconnaissance, and sometimes even attacks.

Based on size:
Small drones: These are lightweight drones that typically fit in the palm of your hand. They are frequently utilized for indoor flying or brief distance flights.

Medium drones: These are larger than small drones and capable of outdoor flying and carrying payloads.

Large drones: These are small aircraft-sized drones with the ability to fly long distances while carrying heavy payloads.

Based on application:
Delivery drones: These are drones that are used for delivering packages or goods to customers.

Inspection drones: These are drones that are used for inspecting infrastructure, such as bridges, buildings, or power lines.

Surveillance drones: These drones are used to observe and gather information about a location or object.

Agricultural drones: These drones are used for monitoring crops and fields, and can assist in crop management.

Based on the power source:
Battery-powered drones: These are drones that depend on battery-powered batteries for their power source.

Gas-powered drones: These are drones that use fuel, such as gasoline or diesel, to power their engines.

Based on flight capabilities:
Fixed-wing drones: These are drones that have a decent wing structure and are intended to fly like a plane.

Rotary-wing drones: These are drones that utilize revolving wings, like rotors or propellers, to accomplish lift and are intended to fly like a helicopter.

Based on data processing:
Edge computing drones: These are drones that process data on-board, without requiring an internet connection or external computing resources.

Cloud computing drones: These drones rely on decentralized computing resources to manage their information and make decisions.

These are just some of the taxonomies that can be used to classify the Internet of Drones. The taxonomy used will depend on the context and purpose of the classification.

One of the biggest challenges to safeguarding the IoD is the lack of regulation and standardization. A universal security architecture is difficult to build since the IoD ecosystem includes so many different parties, such as end-users, network service providers, and drone manufacturers. Additionally, the limited processing and storage capabilities of drones make it challenging to incorporate elaborate security methods. Stakeholders must collaborate to develop IoD security standards and guidelines. Protection of privacy is still another significant issue. Drones that collect and communicate personal data run the risk of data breaches and unauthorized access.

8.2.3 Security threats in drones

Cyber-attacks [8] based on drones have the potential to cause everything from data theft and business disruption to physical damage or injury. The several types of attacks are as follows.

8.2.3.1 Denial of service attack

A DoS attack can overwhelm the drone's organization with meaningless requests, leading the framework to fail and the drone to lose control. Because they prevent the drone and ground controller from interacting, de-authentication attacks allow the attacker to seize control of the drone and use it for malicious purposes.

8.2.3.2 Man-in-the-middle attack

A "man-in-the-middle attack" is a sort of cyber-attack in which an adversary places themself in the path of a drone's communication with a client to intercept and change it. The attacker can link the drone to the Pine-AP SSID (service set identifier) pool and use a device called Wi-Fi Pineapple to monitor all potential paths that the client programming may be taking. This gives them the ability to send the regulator specific commands and impalpably carry out actions on the drone. This attack can be carried out using a weak encryption technique or by a communication chain lacking secure encryption schemes.

8.2.3.3 GPS spoofing

Attackers use a technique called GPS (global positioning system) spoofing to manipulate the GPS signals received by ramblers, causing false location data to be transmitted from the ground controller. Drones that use GPS for navigation can be diverted and taken over using several antennas to send false signals.

8.2.3.4 Jammers attack

Intruders utilize jammers, which are electronic devices, to obstruct signals at the receiver's end. Communication between several users may be challenging as a result. By broadcasting signals on the same frequency as the target and so overriding the frequency signals, jammers interfere with every type of transmission that the target can be configured to.

8.2.3.5 Wormhole attack

Wormhole attacks, which are severe in MANETs, are unavoidable in UAV (unnamed aerial vehicle) networks that use FANET or MANET. In a wormhole attack, the perpetrators carefully position themselves within the network to listen in on the drone network's transmission, record it at one point, and then tunnel the data to a second perpetrator to replay it. This type of re-steering makes it possible for the attacker to attack from any location within the organization, endangering the privacy of sensitive data.

Other types of attacks on drones include logical attacks, malware infection, backdoor access, social engineering, baiting, and message modification [9].

8.2.4 Security techniques to mitigate the threats

8.2.4.1 Machine learning techniques used in drones

Artificial intelligence is essential when trained machines carry out tasks that humans perform. AI is used in various fields, which include wireless communication, automated translation, and speech recognition. In addition, the methods employed for training the machines produce a new framework known as ML, a subset of AI. In this particular instance, ML can respond when a huge number of devices are constantly trying to access network resources for IoT communication. The Internet of Drones uses machine learning techniques to enhance functional areas like security and resource management. ML has offered several solutions for security issues such as GPS spoofing, eavesdropping, jamming, and interference [10]. The machine learning methods and algorithms utilized by drones are shown in the Table 8.1.

8.2.4.2 Fog computing techniques used in drones

A realistic solution for addressing environmental issues is fog computing [16] in the context of UAVs, which also provide energy-efficient data computation and an aerial-to-ground network to promote the quality of service (QoS). To reduce the UAV energy consumption rate, data transmission rate latency, and time computation rate, various fog techniques have been used. Many researchers proposed fog-computing-based models for UAVs, shown in the Table 8.2.

8.2.4.3 Federated learning techniques used in drones

Federated learning (FL) [22] allows AI training at distributed IoT applications without the need for information or data exchange. FL, also referred to as collaborative learning, is the process of training algorithms across several

Table 8.1 Machine-learning-based mitigating techniques.

Ref.	Proposed model	Parameters taken	Techniques used	Advantages	Future work
[11]	OAOFS-MLIDS model (oppositional Aquila optimizer-based feature selection with machine learning enabled intrusion detection system)	NSL-KDD dataset and CICIDS 2017 dataset taken. Performance is evaluated based on accuracy, recall, F-score, and geometric mean	Machine learning algorithms	Recognizes and classifies the type of intrusion in the IOD environment	Advanced deep-learning models will be used to enhance the detection efficiency of the OAOFS-MLIDS model
[12]	Traffic injection detection technique	Monitoring activity in the network. Start and end dates of the flow, source, and destination address, and TCP flags	Machine learning	Outsider attacks and insider attacks can be detected	Only supervised learning is applied. Unsupervised learning will be applied to a swarm of drones
[13]	Hybrid logistic regression and random forest technique	Data are taken from sensors, networks, and drones	Machine learning algorithms	Insider attacks can be detected	Exploration of other attacks like the man-in-the-middle attack and DoS attacks are considered for future work
[14]	UAV- and satellite-based 5G-network security model	CSE-CIC IDS-2018 network dataset taken. Performance is evaluated based on accuracy, recall, F-score, and false negative rate	Machine learning algorithms	Achieves a 99.9% accuracy rate	Further study will be conducted on how technology will incorporate beyond 5G and 6G with machine learning and software defined networks
[15]	Double deep Q-network (DDQN) algorithm	UAV farmland data collection and network data	Machine learning	The intrusion detection system reached a 93.5% accuracy rate for CNN and 94.4% for LSTM	The researcher will develop real-time data capturing to advance the framework for intrusion detection in the future

Table 8.2 Fog computing based mitigating techniques.

Ref.	Proposed model	Parameters taken	Techniques used	Advantages	Future work
[17]	Fog computing aided swarm of drones (FCSD) architecture	Energy consumption rate, latency, reliability	Fog computing	Minimize energy consumption and increase the optimization process	NA
[18]	Global navigation satellite system spoofing detection model	GPS coordinate, UAV velocity. Flight trajectory	Fog computing	More effective than using the IMU sensor alone	A decentralized approach for UAV cluster
[19]	Cloud-fog-edge closed-loop feedback security risk prediction strategy	KDD CUP 99 dataset taken. The deep Boltzmann machine classifies the data and extracts the feature smartly	Fog computing	Provides better accuracy results in the field of network risk prediction. Modularity and interoperability are achieved	NA
[20]	B-drone	Drone-based data are collected	Fog computing	Increases the efficiency, and reduces the network bandwidth, and storage that is needed significantly	File coin-based file storage structure is considered a future work
[21]	Centralized and distributed coordinated scheduling schemes	Time computation rate, energy consumption	Fog computing	Improve the rate of secrecy and energy usage over existing systems	NA

servers or devices using decentralized data samples without transferring the actual data itself. FL, on the other hand, creates more reliable models without sharing data, resulting in solutions with stronger security and access rights to data that protect privacy. The federated learning methods and algorithms utilized by drones are shown in the Table 8.3.

8.2.4.4 Blockchain techniques used in drones

Blockchain technology [28] offers a potentially successful solution for data privacy. To do various jobs, UAVs may need a high data transmission rate, a stable connection, and a high computational resource. Several researchers have developed an algorithm or model based on blockchain that addresses the reliability, availability, and energy efficiency issues. Various mitigation techniques are shown in the Table 8.4.

8.3 Conclusion

UAV networks are rapidly evolving toward the Internet of Drones (IoD) paradigm as it adopts the properties of IoT. In the IoD environment, a network made up of numerous drones transmits and receives data from one another. Due to the lack of security, the possibility and frequency of attacks are very high, which affects the privacy, confidentiality, availability, integrity, and trust of drone data. This book chapter surveys the possible cyber-attacks and the security against such attacks. Various applications and characteristics of IoD are reviewed. The various techniques to enforce security implementing machine-learning-based techniques, fog-computing-based techniques, federated-learning-based techniques, and blockchain-based techniques have been reviewed. One area of anticipated research and development is in the field of IoT security. The Internet of Things (IoT) refers to the interconnected system of physical devices, vehicles, and buildings that collect and exchange data. Drones are a popular application of IoT devices, and their security is a growing concern. Hackers can exploit drones to carry out malicious activities, posing risks to public safety and privacy. Therefore, there is a need for an intelligent anomaly detection system to protect against unauthorized drone activities.

Another area of anticipated research is in the development of security techniques for drones. The use of deep learning and machine learning algorithms can enhance the security of drones by detecting threats and protecting against potential threats. In conclusion, future work in the security techniques of the Internet of Drones involves the development of intelligent threat detection systems and the implementation of advanced security techniques such as deep learning algorithms and encryption methods. These advancements are

Table 8.3 Federated learning based mitigating techniques.

Ref.	Proposed model	Parameters taken	Techniques used	Advantages	Future work
[23]	Multidimensional contract-matching incentive design	Energy consumption, duration taken for traversal and sensing, cost for traversal and sensing, energy transmission	Federated learning, Gale–Shapley algorithm	Privacy is preserved, and communication is made easy here	Researchers have taken into account the adoption of wireless charging techniques such that the UAVs can perform simultaneous model training and intrusion detection
[24]	Power control in secure FL (PCSF) algorithm – to counter eavesdropper	System security rate, energy consumption, FL training time	Federated learning	Low computational complexity, maximizing the security rate, and small federated training time	NA
[25]	Novel federated learning (FL) based fast and fair offloading strategy	Latency, energy consumption, throughput, fairness	Federated learning	It can reduce energy consumption and latency by 23% and 15%, as well as increase throughput and fairness	Investigation of trajectory planning (energy consumption, latency, and QoS)
[26]	Blockchain and FL synergy in drone edge intelligence	Scalability, energy consumption	Federated learning	Privacy is preserved	Channel quality and signal strength, trajectory planning at drones edge Intelligence, drone edge intelligence deployment, resource allocation, obstacles, traffic load, connectivity, and accuracy
[27]	FL-assisted secure data-acquisition scheme	Data transmission rate, execution time	Federated learning	Data privacy is preserved, more feasible	The Internet of Drones embedded with a green energy-harvesting scheme will be investigated in a future study

Table 8.4 Block Chain based mitigating techniques.

Ref.	Proposed model	Parameters taken	Techniques used	Advantages	Future work
[29]	Malicious drone detection model enabled by B5G	Honesty, low coverage distance, and end-to-end delay data transaction	Blockchain	The ability to detect malicious nodes is greatly enhanced	The privacy issue is unfortunately inherent to the non-spurning technique
[30]	Blockchain-based drone delivery system	The number of data transactions, authentication time, and the transaction fee	Blockchain	When the drone is in flight, it is capable of detecting intruders	To increase efficiency, drone trajectory optimization, and path planning can be added
[31]	AKMS-AgriIoT	Data transactions are collected during flying zones	Blockchain	It is possible to counteract various potential attacks in an IoT environment by AKMS-AgriIoT	NA
[32]	Ethereum blockchain model	GPS coordinates, transaction data	Blockchain	It secures UAVs from GPS spoofing using blockchain	When compared to existing approaches, it has relatively minimal risk in terms of secrecy and integrity
[33]	Blockchain-based access control mechanism	Data transactions are collected during flying zones	Blockchain	It can resist various attacks including active and passive attacks	The addition of more useful functionalities such as the ability to keep anonymity when flying drones in IoD environments will also be part of the next improvements for BACSIoD

necessary to protect against potential risks and malicious activities associated with the increasing use of drones. By ensuring the security of drones, we can continue to benefit from their various applications while minimizing potential threats to public safety and privacy.

References

[1] Lagkas, T., Argyriou, V., Bibi, S., and Sarigiannidis, P. "UAV IoT framework views and challenges: Towards protecting drones as 'things,'" Sensors (Switzerland), vol. 18, no. 11. MDPI AG, 2018. doi: 10.3390/s18114015.

[2] Hassanalian M., and Abdelkefi, A. "Classifications, applications, and design challenges of drones: A review," Progress in Aerospace Sciences, vol. 91. Elsevier Ltd, pp. 99–131, May 01, 2017. doi: 10.1016/j.paerosci.2017.04.003.

[3] Hall, R.J. "A geocast-based algorithm for a field common operating picture," in Proceedings - IEEE Military Communications Conference MILCOM, 2012. doi: 10.1109/MILCOM.2012.6415848.

[4] Choudhary, G., Sharma, V., Gupta, T., Kim, J., and You, I. "Internet of Drones (IoD): Threats, Vulnerability, and Security Perspectives," Aug. 2018, [Online]. Available: http://arxiv.org/abs/1808.00203

[5] Abualigah, L., Diabat, A., Sumari, P., and Gandomi, A.H. "Applications, Deployments, and Integration of Internet of Drones (IoD): A Review," IEEE Sensors Journal, vol. 21, no. 22. Institute of Electrical and Electronics Engineers Inc., pp. 25532–25546, Nov. 15, 2021. doi: 10.1109/JSEN.2021.3114266.

[6] Abdelmaboud, A. "The internet of drones: Requirements, taxonomy, recent advances, and challenges of research trends," Sensors, vol. 21, no. 17. MDPI AG, Sep. 01, 2021. doi: 10.3390/s21175718.

[7] Derhab, A. et al., "Internet of drones security: Taxonomies, open issues, and future directions," Vehicular Communications, vol. 39. Elsevier Inc., Feb. 01, 2023. doi: 10.1016/j.vehcom.2022.100552.

[8] Hassija, V. et al., "Fast, Reliable, and Secure Drone Communication: A Comprehensive Survey," IEEE Communications Surveys and Tutorials, vol. 23, no. 4, pp. 2802–2832, 2021, doi: 10.1109/COMST.2021.3097916.

[9] Yaacoub, J.P., Noura, H., Salman, O., and Chehab, A. "Security analysis of drones systems: Attacks, limitations, and recommendations," Internet of Things (Netherlands), vol. 11. Elsevier B.V., Sep. 01, 2020. doi: 10.1016/j.iot.2020.100218.

[10] Bithas, P.S., Michailidis, E.T., Nomikos, N., Vouyioukas, D., and Kanatas, A.G. "A survey on machine-learning techniques for UAV-based communications," Sensors (Switzerland), vol. 19, no. 23. MDPI AG, Dec. 01, 2019. doi: 10.3390/s19235170.

[11] Perumalla, S. Chatterjee, S., and Kumar, A.P.S. "Modelling of oppositional Aquila Optimizer with machine learning enabled secure access control in Internet of drones environment," Theor Comput Sci, vol. 941, pp. 39–54, Jan. 2023, doi: 10.1016/j.tcs.2022.08.019.

[12] Guerber, C., Royer, M., and Larrieu, N. "Machine Learning and Software Defined Network to secure communications in a swarm of drones," Journal of Information Security and Applications, vol. 61, Sep. 2021, doi: 10.1016/j.jisa.2021.102940.

[13] Aldaej, A., Ahanger, T. A., Atiquzzaman, M., Ullah, I., and Yousufudin, M. "Smart Cybersecurity Framework for IoT-Empowered Drones: Machine Learning Perspective," Sensors, vol. 22, no. 7, Apr. 2022, doi: 10.3390/s22072630.

[14] Shrestha, R., Omidkar, A., Roudi, S.A., Abbas, R., and Kim, S. "Machine-learning-enabled intrusion detection system for cellular connected uav networks," Electronics (Switzerland), vol. 10, no. 13, Jul. 2021, doi: 10.3390/electronics10131549.

[15] Fu, R., Ren, X., Li, Y., Wu, Y., Sun, H., and Al-Absi, M.A. "Machine Learning-Based UAV Assisted Agricultural Information Security Architecture and Intrusion Detection," IEEE Internet Things J, 2023, doi: 10.1109/JIOT.2023.3236322.

[16] Gupta, A., and Gupta, S.K. "A survey on green unmanned aerial vehicles-based fog computing: Challenges and future perspective," Transactions on Emerging Telecommunications Technologies, vol. 33, no. 11, Nov. 2022, doi: 10.1002/ett.4603.

[17] Hou, X., Ren, Z., Wang, J., Zheng, S., Cheng, W., and Zhang, H. "Distributed Fog Computing for Latency and Reliability Guaranteed Swarm of Drones," IEEE Access, vol. 8, pp. 7117–7130, 2020, doi: 10.1109/ACCESS.2020.2964073.

[18] He, D., Qiao, Y., Chan, S., and Guizani, N. "Flight Security and Safety of Drones in Airborne Fog Computing Systems," IEEE Communications Magazine, vol. 56, no. 5, pp. 66–71, May 2018, doi: 10.1109/MCOM.2018.1700916.

[19] Li, Q., Tian, Y., Wu, Q., Cao, Q., Shen, H., and Long, H. "A Cloud-Fog-Edge Closed-Loop Feedback Security Risk Prediction Method," IEEE Access, vol. 8, pp. 29004–29020, 2020, doi: 10.1109/ACCESS.2020.2972032.

[20] Khan, A.A. et al., “A drone-based data management and optimization using metaheuristic algorithms and blockchain smart contracts in a secure fog environment,” Computers and Electrical Engineering, vol. 102, Sep. 2022, doi: 10.1016/j.compeleceng.2022.108234.
[21] Al-Abiad, M.S., and Hossain, M.J. “Coordinated Scheduling and Decentralized Federated Learning using Conflict Clustering Graphs in Fog-assisted IoD Networks,” IEEE Trans Veh Technol, Mar. 2022, doi: 10.1109/TVT.2022.3217963.
[22] Nguyen, D.C., Ding, M., Pathirana, P.N., Seneviratne, A., Li, J., and Vincent Poor, H. “Federated Learning for Internet of Things: A Comprehensive Survey,” IEEE Communications Surveys and Tutorials, vol. 23, no. 3. Institute of Electrical and Electronics Engineers Inc., pp. 1622–1658, Jul. 01, 2021. doi: 10.1109/COMST.2021.3075439.
[23] Lim, W.Y. B. et al., “Towards Federated Learning in UAV-Enabled Internet of Vehicles: A Multi-Dimensional Contract-Matching Approach,” IEEE Transactions on Intelligent Transportation Systems, vol. 22, no. 8, pp. 5140–5154, Aug. 2021, doi: 10.1109/TITS.2021.3056341.
[24] Yao, J., and Ansari, N. “Secure Federated Learning by Power Control for Internet of Drones,” IEEE Trans Cogn Commun Netw, vol. 7, no. 4, pp. 1021–1031, Dec. 2021, doi: 10.1109/TCCN.2021.3076167.
[25] Rahbari, D., Alam, M.M., Le Moullec, Y., and Jenihhin, M. “Fast and Fair Computation Offloading Management in a Swarm of Drones Using a Rating-Based Federated Learning Approach,” IEEE Access, vol. 9, pp. 113832–113849, 2021, doi: 10.1109/ACCESS.2021.3104117.
[26] Alsamhi, S.H., et al., “Drones’ Edge Intelligence over Smart Environments in B5G: Blockchain and Federated Learning Synergy,” IEEE Transactions on Green Communications and Networking, vol. 6, no. 1, pp. 295–312, Mar. 2022, doi: 10.1109/TGCN.2021.3132561.
[27] Islam, A., Al Amin, A., and Shin, S.Y. “FBI: A Federated Learning-Based Blockchain-Embedded Data Accumulation Scheme Using Drones for Internet of Things,” IEEE Wireless Communications Letters, vol. 11, no. 5, pp. 972–976, May 2022, doi: 10.1109/LWC.2022.3151873.
[28] Han, T. et al., “Emerging Drone Trends for Blockchain-Based 5G Networks: Open Issues and Future Perspectives,” IEEE Netw, vol. 35, no. 1, pp. 38–43, Mar. 2021, doi: 10.1109/MNET.011.2000151.
[29] Akram, J., Umair, M., Jhaveri, R.H., Riaz, M.N., Chi, H., and Malebary, S. “Chained-Drones: Blockchain-based privacy-preserving framework for secure and intelligent service provisioning in Internet of Drone Things,” Computers and Electrical Engineering, vol. 110, Sep. 2023, doi: 10.1016/j.compeleceng.2023.108772.

[30] Cheema, M.A. et al., "Blockchain-based secure delivery of medical supplies using drones," Computer Networks, vol. 204, Feb. 2022, doi: 10.1016/j.comnet.2021.108706.

[31] Bera, B., Vangala, A., Das, A.K., Lorenz, P., and Khan, M.K. "Private blockchain-envisioned drones-assisted authentication scheme in IoT-enabled agricultural environment," Comput Stand Interfaces, vol. 80, Mar. 2022, doi: 10.1016/j.csi.2021.103567.

[32] Satheesh Kumar, M., Vimal, S., Jhanjhi, N.Z., Dhanabalan, S.S., and Alhumyani, H.A. "Blockchain based peer to peer communication in autonomous drone operation," Energy Reports, vol. 7, pp. 7925–7939, Nov. 2021, doi: 10.1016/j.egyr.2021.08.073.

[33] Bera, B., Chattaraj, D., and Das, A.K. "Designing secure blockchain-based access control scheme in IoT-enabled Internet of Drones deployment," Comput Commun, vol. 153, pp. 229–249, Mar. 2020, doi: 10.1016/j.comcom.2020.02.011.

9

Post-quantum Cryptographic Approach for Cyberspace Security

J.J. Jayakanth[1], Vani Rajasekar[2], and Velliangiri Sarveshwaran[1]

[1]Department of Computational Intelligence, School of Computing, College of Engineering and Technology, SRM Institute of Science and Technology, India
[2]Kongu Engineering College, India

Email: jj.jayakanth@gmail.com; velliangiris@gmail.com; vanikecit@gmail.com

Abstract

The most recent quantum computers are capable of decoding secret encryption keys and opening up the network to attack by solving extremely difficult classical cryptography calculations. In contrast to the billions of years of calculation required by conventional computer equipment, they can nearly instantly solve complex mathematical calculations. The emergence of quantum computing and blockchains are two of the main technical developments both in academic and industry domains that are currently in progress. Researchers support the creation of innovative methods that incorporate encryption technology in the post-quantum computing. A lattice was utilized extensively in cryptography and those techniques were employed in both (a) cryptanalysis by utilizing lattice assumption to crack cryptosystems and (b) cryptography by exploiting challenging lattice problems in computational domain. The issues similar to the quantum attack and the execution of post-quantum signature verification in blockchains are a required scientific implementation in today's world because blockchains have considerably advanced in recent years and learned numerous applications in various fields with the assumption that they will significantly improve their security. This chapter offers the theoretical underpinnings for current research in post-quantum cryptography that aims to integrate safe cryptographic primitives into blockchain technology.

9.1 Introduction

Post-quantum cryptography is an area of cybersecurity that focuses on developing cryptographic algorithms and protocols that are resilient against attacks from quantum computers. Quantum computers have the potential to break many of the widely used cryptographic schemes today, such as RSA and ECC, due to their ability to solve certain mathematical problems much faster than classical computers. As quantum computing technology advances, the need for post-quantum cryptography becomes increasingly important to ensure the security and privacy of data in cyberspace. A recent advance in the science of quantum computing brings the effort to create and use quantum-resistant cryptographic techniques for advancement in classical cryptography. The development of Bitcoin and blockchain technology has attracted increasing interest in recent years as a revolutionary technology that efficiently and robustly enables the level of decentralization needed by contemporary software and services. Blockchain is a shared ledger or decentralized database of all the digital events or transactions that have been carried out and traded among numerous parties [1]. Blockchains are already using the fundamental cryptographic techniques, such as hashing and digital signatures, to authenticate transactions and reach consensus. The majority of diversely used blockchain architecture employs a linked list of blocks, where every block has a hash pointer to the one before it and organizes its data using Merkle trees. Those plans and strategies, however, could not ensure that they will meet any potential future security requirements.

While the globalization of the modern computer civilization tends to make security goals more complex, they also include various security considerations such as privacy-preserving mechanisms and also requiring accountability in widespread applications, in addition to fundamental requirements like tamper resistance and trust [2–4]. Due to the widespread adoption of blockchain technology outside of the financial sector, its security and organizational structure cannot be readily changed. Because of this, the security of blockchains should take into account both current attack vectors and potential future security problems. The gradual increase in the use of Internet of Things (IoT) is increasing so quickly that according to some projections, there will be 75 billion IoT devices in use by 2025. In order to support the IoT devices, it will be necessary to standardize protocols and create the proper designs. The most sophisticated processing operations for these devices are typically performed in centralized servers or cloud computing services on the Internet because they are typically battery-dependent and have limited computational and memory resources. Hash functions, symmetrical cryptography, and public-key encryption algorithms (also known as asymmetric cryptographic systems) are frequently employed to maintain IoT node security. First usable encryption methods were developed in the 1970s, public-key cryptosystems [5]. The computational effort necessary to defeat some asymmetric systems, however, has become easier to reach in light of recent developments in computers and communication, which led to an increase in the suggested minimum key size.

9.2 Cryptographic Primitives

Only a few arithmetic operations can be properly determined in quantum computation. To completely decrypt cryptographic codes, it is necessary to create enough logical qubits; a logical cubit is stabilized always and could be composed of hundreds or even thousands of today's physical qubits. Quantum-resistant cryptographic techniques are widely tested using historical and contemporary data formats. Blockchains essentially rely on the elliptic curve digital signature technology for transactional authentication, which is insufficient to counteract the quantum challenge. It has been demonstrated that the Shor algorithm favors quantum over theoretical computer science [6–8]. More specifically, this chapter examines the implementation of PQC primitives in the major blockchain systems and shows the effects of quantum computing threats on blockchains. The key issues with the most suitable post-quantum cryptosystems for blockchains are discussed in particular. It is important for both scientists and technology to be aware of the quantum computing field and its advancements; this chapter can be used as a reference for the creation of post-quantum blockchains.

9.3 Post-quantum Cryptography

Cryptographic techniques that will continue to offer security even if quantum computers become a realization are known as post-quantum cryptography. More specifically, quantum computing is more effective than classical computing because it makes use of quantum-mechanical processes. The foundation of secure websites, the public key infrastructure, will be vulnerable to quantum computers. A quantum computer with the processing capacity to thwart the HTTPS connection is being developed by researchers at universities and businesses, the NSA, and government agencies [9].Generally, quantum computers function on qubits, which are in a superposition of states,

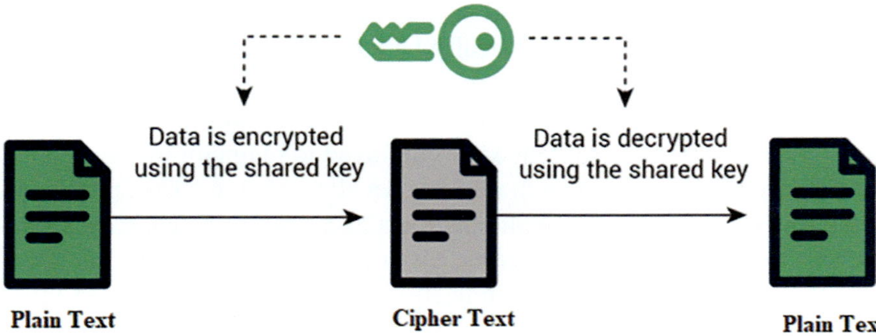

Figure 9.1 Symmetric encryption techniques.

i.e., 0, 1, or both, as opposed to classical computers. They use bits that can only have one of two possible values such as 0 or 1. Because of this, quantum algorithms can use this combination of states to effectively solve a number of mathematical problems for which traditional computers almost never succeed. Although not all issues can be efficiently addressed, a few of them that are now regarded as being tough can be efficiently solved by a quantum computer. Some of these issues serve as the foundation for modern encryption techniques, making them completely unsecure in the post-quantum period [10–11]. The Shor's integer factorization method, which is a quantum algorithm that factors an integer N in polynomial time with regard to the length of N, and the Grover's algorithm, which is a quantum algorithm for searching an unstructured database, are the two most well-known quantum algorithms that directly affect the security of cryptography algorithms. AES-256 and other modern symmetric cyphers with 256-bit keys are thought to be quantum-resistant. In terms of resistance to collisions, hash functions with the appropriate parameters are also regarded as post-quantum secure. The encryption and quantum channel is shown in Figure 9.1 and 9.2 respectively.

As a result, asymmetric algorithms are the main focus of post-quantum cryptography research in order to overwrite RSA, (EC)DH, and (EC)DSA. These post-quantum cryptographic methods depend on supposedly challenging mathematics questions in both the classical and quantum cases [12]. A number of post-quantum digital signature systems exist that rely on various hash functions and its security because they are also post-quantum secure. More specifically, each of the following categories best describes the post-quantum cryptography techniques, each of which bases its security on a unique challenging mathematical equation. The various cryptographic techniques are:

- Code-based cryptography

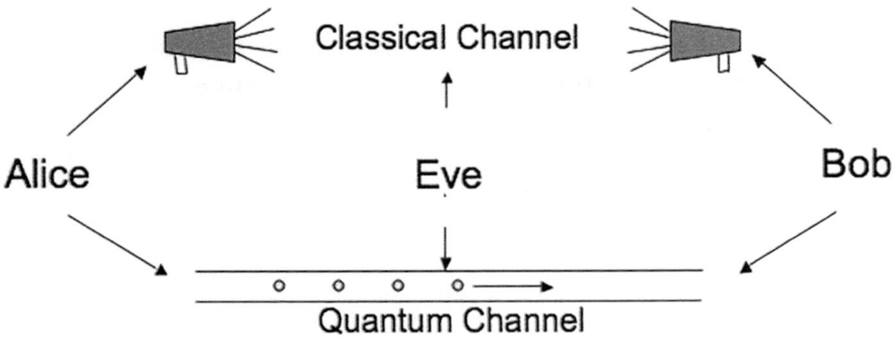

Figure 9.2 Quantum channel.

- Lattice-based cryptography
- Multivariate cryptography
- Hash-based cryptography
- Elliptic curve cryptography
- Hyperelliptic curve cryptography
- Steganography

9.3.1 Code-based cryptography

With the fundamentally different challenge of decoding an incorrect codeword that has been created by an unknown error-correction code, the privacy of the cryptographic methods covered in this class is based on coding theory. The McEliece's cryptosystem, whose privacy is based on the syndrome interpretation problem, is the most traditional example of such a system. Fast encryption and relatively quick decryption are features of McEliece's cryptographic protocol, which is advantageous for carrying out quick blockchain transactions [13]. However, McEliece's cryptosystem necessitates enormous matrices that serve as both the private and public keys, which could be a limitation in contexts with limited resources.

9.3.2 Lattice-based cryptography

This category contains lattice-based cryptographic techniques, which are sets of points in n-dimensional spaces having a periodic structure. The reliability of these algorithms is based on the well-known NP-hardness of particular mathematical problems encountered in the field of lattices, such as the shortest vector problem (SVP), which is concerned with locating the smallest non-zero vector within a lattice. There are other challenging lattice-based challenges as well. At the moment, important possibilities for post-quantum cryptography include lattice-based structures. The well known and most widely used public key schemes are RSA, Diffie-Hellman or Elliptic Curve Cryptographic protocols [14]. Both classical and quantum computers do not seem to be able to break some lattice-based structures. In addition, it is believed that many lattice-based designs are secure under the presumption that several well-known computational lattice issues cannot be addressed effectively.

9.3.3 Multivariate cryptography

Networks of multivariate problems have been shown to be either NP-hard or NP-complete, and multivariate cryptography depends on the difficulty of solving these systems. It is generally accepted that such encryption techniques have some speed restrictions on decryption. It has been established that solving sequences of multivariate polynomial function is NP-complete. Development and cryptanalysis of multivariate cryptography have produced a lot of useful results. Overall, things are more stable now, and the most effective plans have endured the test of time. Since multivariate schemes offer the smallest signature among post-quantum algorithms, it is widely acknowledged that multivariate cryptography proved to be more effective as a method to develop signature schemes.

9.3.4 Hash-based cryptography

This system consists of cryptographic digital signatures that rely on the integrity of the underlying hashing algorithm rather than the difficulty of a mathematical problem for their security. Digital signature techniques like the Merkle signature scheme, zero knowledge, and computation integrity proofs are built using hash-based cryptography. Hash-based signature techniques combine a Merkle tree structure with a Lamport signature or another one-time signature scheme. It is possible to merge numerous one-time signature scheme keys into a single, larger structure because each key can only safely sign one message [15–17]. One-time signature techniques are the foundation of hash-based signature systems. Only one message can be safely signed with a given one-time signing key. Indeed, some of the signing keys are disclosed through signatures. The security of an underlying hash function is the only factor affecting the security of one-time signature methods.

9.3.5 Elliptic curve cryptography

A method of public-key encryption known as elliptic curve cryptography (ECC) is built on the algebraic structure of elliptic curves with finite fields. In contrast to non-EC cryptography, ECC enables fewer keys to offer equal security. Elliptic curves can be used for several applications, including key agreement, cryptographic signature, and pseudo-random generation. By integrating the key agreement with a symmetric encryption method, they can be used for encryption indirect. Numerous elliptic curve-based integer factorization techniques that are utilized in cryptography also make use of elliptic curves [18]. Depending on the intractability of specific mathematical puzzles, public-key cryptography is used. The integrity of early public-key systems was predicated on the idea that it is challenging to factor a huge integer made up of two or more huge prime factors. In order to use elliptic curve cryptography, one must be able to calculate a point multiplication and not be able to do so considering the original and product points.

9.3.6 Hyperelliptic curve cryptography

Hyperelliptic curve cryptography (HECC) is comparable to elliptic curve cryptography (ECC). Because it is an Abelian group, the Jacobian on a hyperelliptic curve can be used as the group in the discrete logarithm issue. The transformation from wired to wireless networks has increased the demand for lightweight cryptography. In view of overcoming the insecurity and resource limitations of wireless systems, numerous symmetric and asymmetric cryptography algorithms have been created. In order to accomplish communication confidentiality, integrity, sender and message authentication, non-repudiation, forward secrecy as well as unforgeability, and public verifiability, signcryption is one such significant development. In contrast to those techniques based on the signature then encryption, signcryption cost is significantly lower because it integrates the signature and encryption.

9.3.7 Steganography

Steganography is the method and technique of concealing information and it is more beneficial in traditional media, electronic content, and numerous social networks. To increase the security of modern wireless multimedia systems and shield data from hackers, many hiding methods are deployed. Existing masking techniques included drawbacks such as larger input sizes, slower processing speeds, and larger keys. As a result, the DNA steganography-based hyperelliptic curve cryptography (HECC), which offers an increased level of protection to image files and ensures the security of digital material, is offered. Steganographic system refers to the amalgamation of embedding and extraction functions for a specific type of cover image. A steganographic system's ability to avoid detection is what makes it secure. Steganalysis is the process of looking for signs of steganography. The aforementioned steganography is utilized to shield the data from unauthorized access. For instance, in the context of safe data transfer, information security has emerged as one of the crucial top concerns [19]. Another age-old method of hiding the information by encoding it is cryptology. However, the need is such that not only must the data being communicated be encrypted, but the knowledge of data transmission must also be hidden. Steganography is a method in which the data are concealed within a carrier in addition to being encoded. Consequently, this offers two levels of security.

9.4 Post-quantum Signature Generation

These cryptographic techniques are expansions of hash-based cryptographic schemes that include desirable cryptographic characteristics of symmetric cyphers in order to create zero-knowledge proofs. They are based on the traditional idea of zero-knowledge proofs. Given that they effectively combine pre-quantum and post-quantum cryptosystems and seek to safeguard traded data against both quantum assaults and attacks against the employed post-quantum schemes, hybrid schemes appear to be the next logical step in post-quantum security. The implementation of two complicated cryptosystems as part of such schemes, however, necessitates a large investment in computer power and increased energy use. As a result, future developers of hybrid post-quantum cryptosystems for blockchains will need to consider how security, high computational, and consumption of resources might be balanced. RSA, digital signature algorithm, and elliptic curve digital signature algorithm are now the most popular cryptographic techniques for digital signatures in practical applications. A digital signature serves as a proof of the originality of a message, communication, or document that was sent electronically. They may be publicly validated by anyone who is aware that it has the maker's signature, much like traditional signature. The three algorithms that make up a digital signature system are as follows:

- Create a key generation approach that produces a public verification key and a private signing key called as keypair.

- Create a signing method called Sign that accepts a message, a private signing key, and produces the message's signature.
- Generate a verification algorithm that takes the message, the public verification key, and the signature and outputs the result that implies whether or not the signature is legitimate.

There are numerous alternatives for quantum safe authentication mechanisms, such as digital signatures: some make use of cryptographic algorithms, others use issues over lattices, and yet others use multi-party computation methods. Key encapsulation mechanisms, or KEMs, could also be utilized in cryptographic approaches to achieve verification.

9.4.1 Hash-based signature generation

The hash-based signature (HBS) technique is strong security-guaranteed, fast, and provides minimal-sized signatures with minimum security procedures. According to the keys and signatures generated, HBS schemes could be classed as stateless or stateful, and then as one-time signature (OTS), few-time signature (FTS), multi-time signature (MTS), or hierarchical signature (HS).

- **Stateful signature scheme:** Typical techniques in this class include the Lamport scheme and its variations. Based on OTS strategies, the private key is ubiquitously created by random mechanism while the public key is adequately inferred from the private key by using a hashing algorithm. Because of the hash function's irreversibility and collision resistance, knowing the public key does not really enable the arithmetic of the private key. Even though the Lamport technique has excellent security features, it is really practically inappropriate because of a number of drawbacks, the first of which is the one-time signature mechanism. Because it is an OTS technique, each secret key can only be used once to sign documents; otherwise, an attacker might be able to create legitimate signatures to obtain information that can be used to impersonate the user. WOTS is a useful option for embedded systems with limited memory because it reduces the space needed for keys and signatures, but at the expense of a slower signature and verifying procedure.
- **Stateless signature scheme:** Stateful hierarchical signature methods have the unique trait that the secret key must be renewed during the authorization process. Keeping track of the utilized one-time keys and ensuring their never-reuse is necessary for stateful signature schemes. SPHINCS installs many HORS-T key pairs and chooses a random one

for each signature generation because stateless techniques do not retain track of used key pairs, ensuring the correct few-time usage of key pairs. A PRNG is used to generate all private keys in stateless systems like the SPHINCS, and one tree is computed in each layer to generate signatures. But stateless systems have the following stability issues. First, because the key pairs are employed in random order rather than successive order, the signature creation is more expensive; thus, various computational strategies that are used in stateful schemes are not appropriate. Although it is believed that post-quantum privacy is present in HBS systems, all possible attack interfaces should be investigated as well. These surfaces primarily come from implementation assaults, such as attacks via side channels and fault attacks. When launching a side channel attack, the attacker observes and/or measures variables such as power usage, electromagnetic leakage, and execution timing to obtain more crucial information.

9.4.2 Post-quantum cryptography in cyberspace resilience

The key aspects of post-quantum cryptography in cyberspace resilience are as follows:

- **Quantum threat:** Quantum computers can efficiently solve problems like integer factorization and discrete logarithms, which form the basis of many encryption methods. This threatens the confidentiality and integrity of data transmitted and stored online. Post-quantum cryptography aims to provide security even in the presence of quantum adversaries.
- **Algorithm development:** Researchers are actively developing and evaluating new cryptographic algorithms that are believed to be secure against quantum attacks. These algorithms include lattice-based cryptography, code-based cryptography, and hash-based cryptography, among others. These new algorithms are designed to be quantum-resistant.
- **Transition period:** The transition to post-quantum cryptography will be challenging, as it involves updating existing systems and protocols. Organizations must develop migration strategies to replace vulnerable cryptographic methods with post-quantum alternatives over time.
- **Standardization:** Several organizations, including NIST (National Institute of Standards and Technology), are working on standardizing post-quantum cryptographic algorithms. These standards will provide

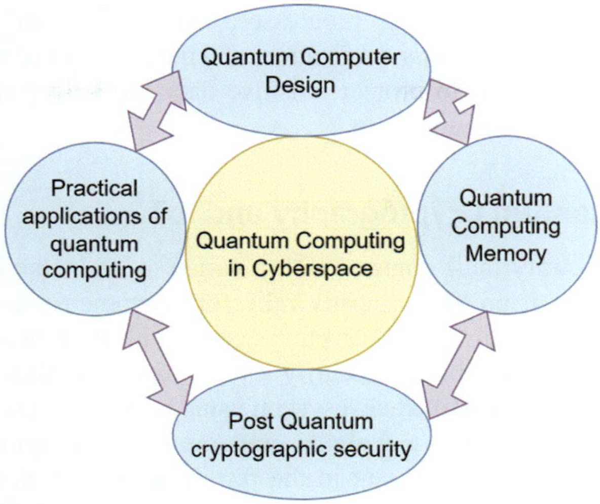

Figure 9.3 Post-quantum cryptography in cyberspace.

guidance to organizations on which algorithms to use for different purposes.

- **Hybrid approaches:** In the short term, hybrid approaches can be employed to combine classical and post-quantum cryptography. This provides a measure of security against quantum attacks while maintaining compatibility with existing systems.
- **Quantum key distribution (QKD):** Quantum key distribution is a quantum technology that can enhance the security of encryption keys by leveraging the principles of quantum mechanics. QKD can be used to create encryption keys that are theoretically secure against quantum attacks.
- **Research and development:** Continuous research and development in post-quantum cryptography are crucial to stay ahead of potential quantum threats. It is important for organizations to stay informed about the latest advancements in the field.
- **Quantum-resistant protocols:** Beyond encryption, other cryptographic protocols and security mechanisms, such as digital signatures, must also be made quantum-resistant.

The post-quantum cryptography plays a crucial role in making cyberspace more resilient to the threats posed by quantum computers. The various post-quantum cryptographies in cyberspace are shown in Figure 9.3. As quantum computing technology progresses, organizations and governments need to invest in research, standardization, and the adoption of post-quantum cryptographic solutions to protect sensitive data and ensure the long-term security of digital communication.

9.5 Pre-quantum Cryptography and IoT

Cyberspace security must contend with a variety of vulnerabilities, but for communications, it presently mostly relies on asymmetric and symmetric cryptographic protocols. Such cryptosystems' robustness has historically been measured by their bits-of-security level, which is a gauge of the computational power needed to crack a system using brute force on conventional computers. Cryptographic techniques and pre-quantum symmetric algorithms appear to still be relevant in the post-quantum epoch because it is thought improbable that effective quantum algorithms would be discovered for NP-hard issues. In fact, it is anticipated that cryptographic encryption and hash functions will just need to increase the size of their key output. A public key is used to encrypt messages addressed to a specific user, and that person uses a private key to decrypt those messages. This is known as public-key encryption. Since a public key and a paired private key are similar in many aspects, the computing effort needed to launch a brute-force key search attack to identify a private key from its associated public key determines the robustness of a public-key cryptosystem [20]. Therefore, public-key encryption depends on mathematical conundrums like discrete logarithms, elliptic curves, and integer factoring for which there was, until recently, no effective answer. The primary benefit of public-key cryptosystems in an unsecured network is that they address the issue of key distribution; since they are asymmetric, the public key prevents message decryption. As symmetric cryptography uses the same key for both message encryption and decryption, it necessitates secure methods for distributing keys among peers who are involved in information exchange. However, a public-key cryptosystem requires its keys to have a specific structure.

9.5.1 Cryptosystems in quantum computing

RSA, ECDSA, elliptic curve DH (ECDH), digital signature algorithm (DSA), and other public-key techniques based on the integer factorization issue, the elliptic curve discrete logarithm problem, or the discrete logarithm problem are all susceptible to quantum assaults. On a suitably powerful quantum computer, Shor's algorithm can quickly solve each of them. It should be highlighted that current 80-bit security cryptographic algorithms can be cracked for anywhere between tens of thousands and hundreds of millions of dollars, but 112-bit security systems are predicted to be resistant to traditional computer attacks for the next 30–40 years. Therefore, it is seen more necessary to move to post-quantum cryptosystems that also withstand classical computer attacks than to improve existing cryptosystems because computationally strong quantum computers are not anticipated to be accessible in the next 20 years. Due to this, some academics have already looked into the security danger posed by quantum computers in various domains in recent years.

9.5.2 Post-quantum cryptosystems

The idea underlying error-correction codes, which have long provided resilience to digital communications, is the basic foundation for code-based cryptosystems. Although McEliece's cryptosystem uses enormous matrices as its private keys and public keys, which is a major downside when using it in resource-constrained IoT devices, it is extremely fast for encryption and relatively fast for decryption. Different compression/decompression methods may be evaluated to address this problem, and various McEliece scheme iterations based on other codes may be suggested [21]. Code-based signature techniques are also available. For IoT advancements, it must be considered that, although the created signatures are brief and can be validated quickly, the needed key size is quite high and signature generation is inefficient in the case of the CFS variations. Additionally, the creation of IoT signature schemes resulting from the use of the Fiat–Shamir transformations on authentication protocols should be taken into account.

9.5.3 Post-quantum threat

The threat to existing protocols will change depending on the availability of completely error-corrected quantum computers. The widely used RSA or elliptical curve methods as well as other asymmetric security protocols can all be decrypted by these quantum systems. These protocols are typically used to transmit secure communications or keys over open networks like the internet. Anyone can encrypt a communication using these protocols, but only the sender has the key to decrypt it. The ability to decode an encrypted message without this key, however, will be enabled by quantum computing and make encrypted messages readable. In traditional security assessments, understanding the value of cybersecurity for a company is normally the first stage, both generally and for the protection of particular systems and products. Making the right plans to defend against threats and react to possible breaches of security involve measuring the worth exposed to these risks. Given the lack of fully error-corrected quantum computers, the first consideration in quantum security measures is timing.

9.5.4 Lattice-based post-quantum cryptography

Lattice-based cryptographic algorithms are built on sets of points called lattices, which are periodic structures in n-dimensional spaces. The use of lattices in telecommunications is related to the alleged difficulty of lattice issues like the quantum algorithms for the shortest independent vector problem (SIVP), closest vector problem (CVP), and shortest vector problem (SVP). Taking the SVP as an example, the goal is to identify the shortest nonzero vectors within the lattice, which is an NP-hard issue that cannot presently be addressed by a quantum method. Strong security proofs are provided by lattice-based cryptosystems, and they are typically easy to use, quick, and efficient to develop. Post-quantum cryptography in cyberspace-based cryptographic protocols for the Internet of Things, however, should be able to manage the collection and maintenance of enormous keys and massive cipher text overhead expenses effectively. For instance, the research and innovation of IoT-optimized lattice-based cryptographic protocols based on polynomial algebra certainly appears to be quite fascinating. The size of the key in the aforementioned lattice-based schemes are longer than the standard dimension of a pre-quantum cryptosystem; despite the fact that it is smaller than those used by code-based or multivariate public-key cryptosystems, conventional methods and optimizations must be used when using them in IoT applications.

9.6 DER (Distributed Energy Sources) Network

As carbon-neutral policies are promoted globally, distributed energy resources (DER) like solar panels and wind turbines are being added to the electricity grid. The percentage of PV and wind turbine electricity production has increased during the last 10 years. In California, solar energy produced 21% of the state's total electricity [22]. The necessity for a network to control the DER system has grown due to the growing number of DERs and their importance to the system's stability. The time it takes to crack current encryption techniques can be significantly reduced by quantum processors' strong calculation abilities. According to predictions, quantum computing assaults will render half of the currently used encryption techniques ineffective. To defend the DER network against upcoming quantum attacks, it is

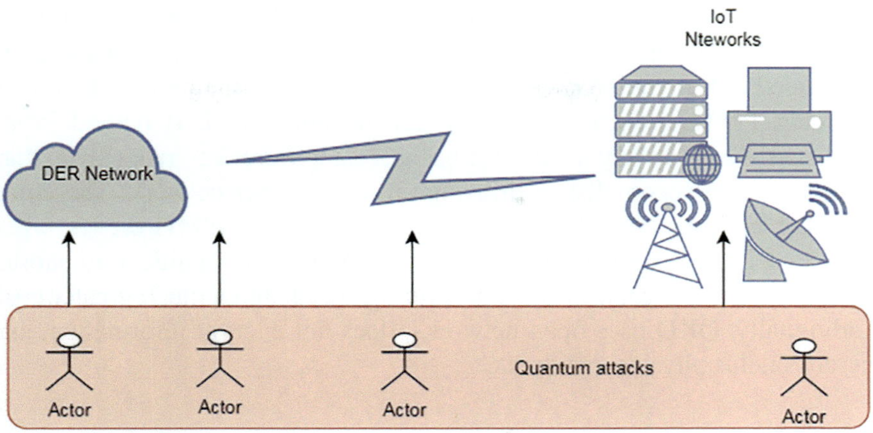

Figure 9.4 Quantum computing attacks.

necessary to anticipate future quantum threats and identify the weaknesses of the DER network. Additionally, new cryptography and security methods should be properly implemented on the DER activities in accordance with the vulnerability. For the stable functioning of the whole system as well as DER systems, DER networks are vast networks that connect DER devices and control devices (SCADA and DERMS) and power plants. The DER network also combines a sensor network with an established data communication network to monitor the DER device's current condition and to regulate it accordingly. Numerous standards have been developed depending on the type of DER system in order to protect the DER network.

9.6.1 Quantum key distribution

A procedure called QKD is used to distribute encryption keys across the quantum channel (Q-channel). Theoretically, it is demonstrated that quantum threats cannot perform cryptanalysis on ciphertext if the cryptographic keys are totally secret and random. Using the principles of quantum physics, the Q-channel could identify whether the broadcast signal has been intercepted or not. Various information like IP addresses are released when an attack gains control over the packet header using sniffing tools like wire-shark or packet capture since Wi-Fi, Raspberry Pi, or Modbus cannot detect whether eavesdropping is taking place. The attacker cannot access any network data since the QKD network can detect eavesdropping. The network halts data transfer when the attacker is able to eavesdrop to prevent information leaking. Thus, using QKD on the DER network can stop intruders from collecting the *r* number of communications needed to overcome AEAD, unlike traditional networks like Wi-Fi, Raspberry Pi, or Modbus. Depending on how the Q-bit is encoded, QKD can be divided into two categories: the prepare and measurement (PM) approach and the quantum entangle base (EB) method. When the distance is relatively small, PM has a greater secret key rate (SKR) than EB and can be accomplished utilizing an optical fiber network. Therefore, the DER network typically uses the PM-based approach. However, when compared to the existing public network, QKD has a significantly shorter communication range, a slower transmitting speed, and a much greater cost. Additionally, QKD uses open networks, does not encrypt information, and merely shields physical networks.

9.6.2 Bitcoin

The most well-known and original blockchain technology is Bitcoin. It does away with the requirement for a centralized authority by being developed as a peer-to-peer means of payment. Bitcoins are supported by cryptographic techniques that let various nodes in network authenticate blockchain transactions in a trust-free setting and preserve them in an immutable, secure ledger. Hashcash serves as the PoW algorithm for Bitcoin [23]. Hashcash was first intended to protect email systems from denial-of-service attacks. To do this, a computationally challenging task had to be solved by the prospective sender before they could send an email. There are two net effects of using the Hashcash PoW method. One reason is because the miners are encouraged to employ specialized hardware, such as ASIC miners, or join mining pools where the effort and reward are shared among many individuals. PoW eliminates incentives for trying to introduce malicious blocks into networks. Due to error detection and correction, these transactions have a very high likelihood of being discarded by the network, guaranteeing very large wasted costs for the prospective miner. The ECDSA signature algorithm is used by the Bitcoin transaction system to confirm authorization and control of tokens and also to provide incontrovertible proof that tokens were spent on transactions that will not be tampered. Secp256-k1 is the elliptic curve used by Bitcoin for its ECDSA. Discovering the private key when the public key is disclosed after a signed transaction has been broadcasted to the network would constitute an efficient quantum attack. To use the private key to sign a brand-new transaction would enable an attacker to pretend to be the key owner. Since Ethereum's authentication system depends on difficulty of the discrete logarithm approach, it is very vulnerable to threats utilizing Shor's algorithm. One slight benefit Ethereum does have over Bitcoin is that it processes transactions much more quickly. However, one significant drawback of Ethereum's use of account-based payments serves as a counterbalance. The private key for the account must be used to sign every single departing transaction, and the key pair must be used to verify the signature.

9.6.3 Litecoin

A source-code fork of the Bitcoin ecosystem created Litecoin, which denotes both Litecoin and Bitcoin have a lot in common. The block time and the PoW method are just two notable examples of how Litecoin differs from other cryptocurrencies. In terms of use cases for electronic payments, it is extremely comparable to Bitcoin. However, it aims to execute transactions more quickly than Bitcoin because of a lower block time. Scrypt is a unique PoW algorithm used more by Litecoin than by Bitcoin. It pursues the similar objective of using computational power to resolve a problem and granting permission to add the subsequent blocks to the blockchain. Scrypt is intended to consume substantially less mining power than Bitcoin, which has a hashing rate of about 46,000,000 TH/s compared to Litecoin's 298 TH/s. The ECDSA technique is used by Litecoin to validate transactions. It uses the secp-256k1 elliptic curve to construct its signature mechanism, just like Bitcoin. Scrypt is possibly open to a quantum 51% attack utilizing Grover's algorithm, just like other PoW systems. For a quantum attacker, this is perhaps the most lucrative attack. When compared to Bitcoin, Litecoin offers the benefit of a faster bandwidth and a minimal waiting time.

9.6.4 Ethereum

Ethereum is a decentralized, freely available blockchain that facilitates smart contracts. Ethereum is the native cryptocurrency of the entire system. After Bitcoin, Ether is the only cryptocurrency with a higher value in the marketplace. On Ethereum, anybody can publish permanent, immutable decentralized apps that let users interact with them. Decentralized finance (DeFi) services allow a variety of financial products and services without a requirement for conventional banking intermediaries like banks, brokerages, or exchanges. For example, they let Bitcoin users borrow against their assets or lend them out for profit. In addition, NFTs – unique tokens recognized by many organizations as proof of ownership of a particular asset or privilege – can be created and traded by Ethereum users [24–25]. The primary goal of the update is to boost the network's transaction throughput from its present rate of roughly 15 transactions per second to potentially upwards of tens of thousands of transactions per second. This will be achieved by dividing the workload among numerous parallel blockchains and making them all share a single consensus proof-of-stake blockchain.

9.6.5 Monero

Monero is a decentralized cryptocurrency. It achieves anonymity and decentralization by utilizing a public distributed ledger with confidentiality algorithms that obscure transactions. Addresses, transaction amounts, address balances, or transaction histories cannot be interpreted by observers. This idea was adopted by the cryptography community to create Monero, which launched its main net in 2014. To hide transaction data, Monero employs ring signatures, zero-knowledge proofs, "stealth addresses," and techniques to mask IP addresses. These functions are built into the protocol, although users have the option of disclosing view keys for external auditing. Among cryptocurrencies, Monero has the third-largest developer community after Ethereum and Bitcoin. Cypherpunks and consumers looking for privacy features not offered by other crypto currencies have been drawn to it by its privacy characteristics. It is increasingly utilized in criminal activity like malware, money laundering, darknet markets, and cryptojacking. The privacy- and anonymity-related aspects of Monero are its main benefits. All transaction data are obscured despite the ledger being open and decentralized. Bitcoin, in comparison, allows for complete transparency and openness regarding all transactional information, user addresses, and wallet balances. Monero has a devoted following among crypto-anarchists, cypherpunks, and privacy advocates as a result of these qualities. The blockchain quantum attacks are specified in Table 9.1.

9.7 Conclusion

A variety of key industries, including remote monitoring, smart cities, precision farming, smart clothing, and Industry 4.0 smart manufacturing, whose applications significantly rely on source of energy and battery-dependent IoT devices, can be enhanced by successful post-quantum IoT cryptographic algorithms. Although there has been a huge increase in interest in post-quantum cryptography over the past few years, much more study needs to be done before it can be used to the Internet of Things. This chapter gave a comprehensive overview of the many factors affecting post-quantum IoT advancement. There is a need to move beyond pre-quantum safety. The most important standards and projects, the most recent post-quantum public-key

Table 9.1 Quantum attacks and its categories in blockchain.

Techniques	Vulnerability level	Attack categories
Bitcoin	High	• Specifically with relation to their signature technique, which are broadcast to network that are susceptible to a quantum attack. • The primary attack type found targets network-declared transactions that have not yet been added to a block. An attacker can discover the private key by utilizing the public key disclosed by the transaction's sender. They will be able to repeat the transaction with any desired output location thanks to this.
Litecoin	High	• Litecoin is similarly susceptible to a quantum attack as Bitcoin because they both share a large portion of their technical architecture. • Similar to Bitcoin, attacks against transactions that have not yet been added to the blockchain but have been reported to the networks are the most harmful.
Ethereum	High	• Reusing public keys is prevalent in account-based systems like the one used by Ethereum. With some Ether tokens left in the account, the threat method that finds potentially target accounts has already reported transactions to the network. • A quantum attacker might create a legitimate transaction signature to falsify transactions in a user's name by utilizing Shor's approach to solve the public key to obtain the private key.
Monero	Medium	• The Monero EdDSA signature method is vulnerable to quantum assaults since it depends on the discrete logarithm problem. The anonymity of its users and the quantities being transferred, however, give Monero considerable resistance to quantum attacks. • Despite the vulnerability of the Monero Bulletproof method, which enables the concealment of transaction quantities, an attacker would have to rely on chance to select a transaction with significant value. • As a result of a recent modification to Monero's consensus mechanism, RandomX, it would also be more resistant against quantum attacks that aim to accomplish a 51% attack using Grover's technique.

cryptosystems for possible IoT applications, and the essential trends and issues that will develop in the coming years are all discussed. Due to the restrictions of the maintenance and the synchronous updating, a large distributed network like the blockchain demands unusual thought while transitioning to a post-quantum cryptography. Such changes necessitate gradual rollouts and rollbacks, in addition to performance validation and application frameworks.

References

[1] Alagic, G., Alperin-Sheriff, J., Apon, D., Cooper, D., Dang, Q., Kelsey, J., ... and Smith-Tone, D. (2020). Status report on the second round of the NIST post-quantum cryptography standardization process. US Department of Commerce, NIST.

[2] Alagic, G., Alagic, G., Alperin-Sheriff, J., Apon, D., Cooper, D., Dang, Q., ... and Smith-Tone, D. (2019). Status report on the first round of the NIST post-quantum cryptography standardization process (pp. 052419-7). Washington, DC: US Department of Commerce, National Institute of Standards and Technology.

[3] Zeydan, E., Turk, Y., Aksoy, B., and Ozturk, S. B. (2022, February). Recent Advances in Post-Quantum Cryptography for Networks: A Survey. In 2022 Seventh International Conference On Mobile And Secure Services (MobiSecServ) (pp. 1–8). IEEE.

[4] Dhanaraj, R. K., Rajasekar, V., Islam, S. H., Balusamy, B., and Hsu, C. H. (Eds.). (2022). Quantum Blockchain: An Emerging Cryptographic Paradigm. John Wiley & Sons.

[5] Rajasekar, V., Krishnamoorthi, S., Saračević, M., Pepic, D., Zajmovic, M., and Zogic, H. (2022). Ensemble Machine Learning Methods to Predict the Balancing of Ayurvedic Constituents in the Human Body: Ensemble Machine Learning Methods to Predict. Computer Science, 23(1).

[6] Joseph, D., Misoczki, R., Manzano, M., Tricot, J., Pinuaga, F. D., Lacombe, O., ... and Hansen, R. (2022). Transitioning organizations to post-quantum cryptography. Nature, 605(7909), 237-243.

[7] Dhanaraj, R.K., Islam, S.H., and Rajasekar, V. A cryptographic paradigm to detect and mitigate blackhole attack in VANET environments. Wireless Netw (2022). https://doi.org/10.1007/s11276-022-03017-6

[8] Rajasekar, V., Predić, B., Saracevic, M., Elhoseny, M., Karabasevic, D., Stanujkic, D., and Jayapaul, P. (2022). Enhanced multimodal biometric recognition approach for smart cities based on an optimized fuzzy genetic algorithm. Scientific Reports, 12(1), 1–11.

[9] Borges, F., Reis, P. R., and Pereira, D. (2020). A comparison of security and its performance for key agreements in post-quantum cryptography. IEEE Access, 8, 142413-142422.

[10] Tian, J., Lin, J., and Wang, Z. (2019, October). Ultra-fast modular multiplication implementation for isogeny-based post-quantum cryptography. In 2019 IEEE International Workshop on Signal Processing Systems (SiPS) (pp. 97–102). IEEE.

[11] Velliangiri, S., Manoharn, R., Ramachandran, S., Krishnasamy, V., Rajasekar, V. R., Karthikeyan, P., ... and Dhanabalan, S. S. (2021). An efficient lightweight privacy preserving mechanism for industry 4.0 based on elliptic curve cryptography. IEEE Transactions on Industrial Informatics.

[12] Hasija, T., Ramkumar, K. R., Kaur, A., Mittal, S., and Singh, B. (2022, June). A Survey on NIST Selected Third Round Candidates for Post Quantum Cryptography. In 2022 7th International Conference on Communication and Electronics Systems (ICCES) (pp. 737–743). IEEE.

[13] Rajasekar, V., Venu, K., Jena, S. R., Varthini, R. J., and Ishwarya, S. (2021). Detection of Cotton Plant Diseases Using Deep Transfer Learning. Journal of Mobile Multimedia, 18(2), 307–324.

[14] Señor, J., Portilla, J., and Mujica, G. (2022). Analysis of the NTRU Post-Quantum Cryptographic Scheme in Constrained IoT Edge Devices. IEEE Internet of Things Journal.

[15] Guo, Q., Johansson, T., and Nilsson, A. (2020, August). A key-recovery timing attack on post-quantum primitives using the Fujisaki-Okamoto transformation and its application on FrodoKEM. In Annual International Cryptology Conference (pp. 359–386). Springer, Cham.

[16] Rajasekar, V., Sathya K., and Premalatha, J. “Energy Efficient Cluster Formation in Wireless Sensor Networks Based on Multi Objective Bat Algorithm,” 2018 International Conference on Intelligent Computing and Communication for Smart World (I2C2SW), 2018, pp. 116–120, doi: 10.1109/I2C2SW45816.2018.8997529.

[17] Paquin, C., Stebila, D., and Tamvada, G. (2020, April). Benchmarking post-quantum cryptography in TLS. In International Conference on Post-Quantum Cryptography (pp. 72–91). Springer, Cham.

[18] Aleksandrova, E. B., Shtyrkina, A. A., and Iarmak, A. V. (2019). Post-quantum primitives in information security. Nonlinear Phenomena in Complex Systems, 22(3), 269–276.

[19] Sathya, K., Premalatha, J., and Rajasekar, V. (2015, December). Random number generation based on sensor with decimation method. In 2015

IEEE Workshop on Computational Intelligence: Theories, Applications and Future Directions (WCI) (pp. 1–5). IEEE.

[20] Rajasekar, V., Varadhaganapathy, S., Sathya, K., and Premalatha, J. (2016, March). An efficient lightweight cryptographic scheme of signcryption based on hyperelliptic curve. In 2016 3rd International Conference on Recent Advances in Information Technology (RAIT) (pp. 394–397). IEEE.

[21] Bos, J. W., Costello, C., Naehrig, M., and Stebila, D. (2015, May). Post-quantum key exchange for the TLS protocol from the ring learning with errors problem. In 2015 IEEE Symposium on Security and Privacy (pp. 553–570). IEEE.

[22] Nejatollahi, H., Dutt, N., Ray, S., Regazzoni, F., Banerjee, I., and Cammarota, R. (2019). Post-quantum lattice-based cryptography implementations: A survey. ACM Computing Surveys (CSUR), 51(6), 1–41.

[23] Rajasekar,V.,Jayapaul,P.,Krishnamoorthi,S.,Saracevic,M.,Elhoseny,M., and Al-Akaidi, M. (2021). Enhanced WSN Routing Protocol for Internet of Things to Process Multimedia Big Data. Wireless Personal Communications, 1–20.

[24] Joseph, D., Misoczki, R., Manzano, M., Tricot, J., Pinuaga, F. D., Lacombe, O., ... and Hansen, R. (2022). Transitioning organizations to post-quantum cryptography. Nature, 605(7909), 237–243.

[25] Borges, F., Reis, P. R., and Pereira, D. (2020). A comparison of security and its performance for key agreements in post-quantum cryptography. IEEE Access, 8, 142413–142422.

10

Navigating the Boundless: Security in Cyberspace and Outer Space

R. Mahalakshmi and Minakshi Gautam

Presidency University, India

Email: minakshi2022cse0009@presidencyuniversity.in; mahalakshmi@presidencyuniversity.in

Abstract

In the chapter "Navigating the Boundless: Security in Cyberspace and Outer Space," the terms "cyberspace" and "outer space" refer to two critical domains that are increasingly important to modern society. Cyberspace refers to the interconnected digital networks that comprise the Internet and other communication systems. In contrast, outer space refers to the area beyond Earth's atmosphere where satellites and other spacecraft operate. Both domains are vulnerable to various threats, including cyber-attacks, space debris, and other security concerns. This chapter comprehensively explores the challenges and complexities in ensuring security in two critical domains. The literature review section provides an overview of the current state of security measures in cyberspace and outer space, including the various types of threats and the policies and regulations to address them. The chapter then discusses the methodologies used to address security concerns in these domains, including encryption techniques, firewalls, and intrusion detection systems. Continuous monitoring and vulnerability assessments are essential to ensure that security systems remain up-to-date and effective. The applications section of the chapter explores how these security methodologies are implemented, including using encryption in secure satellite communication systems firewalls, and intrusion detection systems to protect against cyber-attacks. The section also discusses the challenges of applying these methodologies in real-world situations, such as the limited resources available for satellite-based systems.

Finally, the conclusion emphasizes the importance of continued research and development to address the ever-evolving threats to security in cyberspace and outer space. The section underscores the need for collaboration and international cooperation to ensure these domains remain secure and stable in the face of emerging threats. Overall, the chapter thoroughly examines the complexities of ensuring security in these two boundless domains.

10.1 Introduction

People have been curious about the skies above them for thousands of years and have developed myths and hypotheses about the universe. While telescopes made significant progress in our understanding, the late 1950s saw a revolution when people could launch artificial satellites into Earth's orbit. Since then, various technological and space races have occurred between nation-states, some of which have resulted in significant international cooperation (like the International Space Station), and others have caused international tensions and technological failure.

The term cyberspace refers to the virtual world of computers and the Internet. On the other hand, outer space refers to the physical universe beyond Earth's atmosphere. Both cyberspace and outer space have become critical components of modern society, and their security is paramount. This literature review will explore the concepts of cyberspace and outer space security and the various issues surrounding them.

10.1.1 Cyberspace

Cyberspace was first introduced in William Gibson's book *Necromancer*, where it was portrayed as a virtual world created by a network of computers. The term became popular and was included in many English dictionaries. Cyberspace is a virtual environment people use to communicate over computer networks. It is a limitless space with no physical boundaries, and it is made up of bits and bytes that are constantly changing. Cyberspace is an imaginary place where two parties can interact. The need for computer security has grown in the 21st century because cybersecurity, also called information technology security or electronic information security, encompasses the strategies and processes employed to safeguard computers, networks, databases, and applications from unauthorized access and attacks. With the rise of cybercrime, digital currency, and e-governance, cybersecurity specialists and forensic experts face increasing threats. Web applications, the primary platform for delivering services and information over the Internet, are particularly vulnerable. Threat intelligence, big data, and machine learning techniques are crucial in effectively detecting, analyzing, and defending against these threats in real time. Cybersecurity entails protecting electronic systems, networks, and data from malicious activities and is a vital practice in our technologically driven world.

10.1.1.1 Cyberspace and laws

The legal framework that governs the use of the Internet and technology ensures that individuals and organizations are held accountable for their actions online. Cyber laws protect people's privacy, data, and intellectual property rights and prevent cybercrime and other malicious activities. These laws apply to online activities, including e-commerce, social media, online banking, and other digital transactions. Cyber laws aim to promote a safe and secure digital environment for all users while promoting innovation and growth in the digital economy. Cyber laws establish rules and regulations for using the Internet and technology, holding individuals and organizations responsible for their online actions. The primary objective of these laws is to safeguard people's privacy, data, and intellectual property rights and to prevent cybercrime and other harmful online activities. These laws apply to various digital transactions, such as e-commerce, social media, and online banking, and aim to create a secure online environment that fosters innovation and growth in the digital economy. Cyber laws deal with legal issues related to communication, distribution, and transactions involving information devices and technologies connected to a network. Unlike property law, it covers various laws and regulations, including statutory, legal, and constitutional provisions concerning computers and the Internet. Cyber laws apply to individuals and institutions that provide access to cyberspace, develop software/hardware to enter cyberspace, or use their computer systems to access it.

Essentially, any citizen's activities related to legal aspects of cyberspace fall under the scope of cyber laws. The two primary acts defining different cybersecurity aspects in India are the Indian Penal Code of 1860 and the Information Technology Act of 2000. Cyberspace is an electronic medium designed to enable online communication worldwide through large computer networks, including sub-networks that follow the TCP/IP protocol. TCP is a communication standard that ensures successful data transmission across the Internet, while IP is a method of sending data from one device to another using the Internet. Each device has a unique IP address that identifies it and enables communication and data exchange with other devices across the Internet. In cyberspace, users share information, interact, and engage in various activities such as social media, as described by William Gibson in his

1984 novel, "Neuromancer." The concept of cyberspace continues to evolve and is widely used for different purposes.

Cyber laws are crucial in addressing and regulating various cybersecurity and digital activities. Here are three case studies to illustrate the practical implications of cyber laws:

United States v. Ross Ulbricht (Silk Road Case):

Case overview: The Silk Road was an infamous online marketplace on the dark web that facilitated the anonymous sale of illegal drugs, weapons, and other contraband using cryptocurrencies. Ross Ulbricht, operating under the pseudonym "Dread Pirate Roberts," was the alleged mastermind behind Silk Road.

Cyber law implications: Ulbricht was arrested and charged with multiple offenses, including money laundering, computer hacking, and conspiracy to commit narcotics trafficking. His prosecution highlighted the ability of law enforcement agencies to trace and apprehend individuals involved in illicit online activities. It also underscored the importance of cyber laws in addressing crimes involving cryptocurrencies and dark web marketplaces.

Facebook-Cambridge Analytica Data Scandal:

Case overview: In 2018, it was revealed that the British political consulting firm Cambridge Analytica had harvested data from millions of Facebook users without their consent. This data was used for political advertising and manipulation during the 2016 US presidential election.

Cyber law implications: The incident led to investigations by US and European authorities, highlighting the importance of data privacy and protection laws. In response to the scandal, regulations like the General Data Protection Regulation (GDPR) in Europe and various data protection laws in the United States were strengthened or enacted. Companies are now held more accountable for data breaches and misuse of personal information.

Stuxnet Worm and International Cyber Conflict:

Case overview: Stuxnet is a computer worm discovered in 2010 and believed to have been developed by the US and Israeli governments. It targeted Iran's nuclear facilities, specifically the centrifuges used for uranium enrichment. Stuxnet was designed to sabotage these facilities by altering the operation of industrial control systems.

Cyber law implications: Stuxnet raised complex legal and ethical questions surrounding cyber warfare and state-sponsored cyber-attacks. It illustrated the need for international agreements and norms related to cyber conflict. It also prompted discussions about the attribution of cyber-attacks and the potential consequences for states engaging in such activities.

These case studies demonstrate the real-world impact of cyber laws in addressing cybercrimes, data breaches, and international cyber conflicts. They underscore the importance of legal frameworks, regulations, and international cooperation in the digital age. Cyber laws are essential in ensuring accountability, protecting individual rights, and maintaining digital systems and data security and integrity.

10.1.1.2 Cyberspace security

Cyberspace security, or cybersecurity, is essential for safeguarding computer systems and networks against unauthorized access, theft, and damage to hardware, software, and data. The advent of digital technology has resulted in a significant surge in cyber-attacks, highlighting the criticality of cybersecurity. One of the major obstacles in this field is the ever-changing landscape of technology, which continuously introduces new vulnerabilities and attack vectors. Cybercriminals and hackers innovate and devise novel techniques to breach security systems and pilfer data. As a result, robust security measures like encryption, firewalls, and intrusion detection systems have become imperative. Another challenge in the realm of cybersecurity is the proliferation of interconnected devices. The Internet of Things (IoT) has fostered an extensive network of linked devices, amplifying vulnerability to cyber-attacks. As the number of connected devices grows, so does the potential for cyber-attacks. Governments and organizations worldwide invest heavily in cybersecurity to protect their digital assets. They are also collaborating to develop global standards for cybersecurity to ensure that all parties are protected from cyber threats. Figure 10.1 shows the increase in cyber breaches reported till 2022.

10.1.1.3 Challenges and threats

Cyberspace faces many challenges and threats, many of which are constantly evolving as technology advances. Some specific challenges and threats in cyberspace include the following.

Cyber-attacks: These can take various forms, including malware, ransomware, distributed denial-of-service (DDoS) attacks, and phishing. Cybercriminals and state-sponsored actors use these methods to steal data, disrupt services, or gain unauthorized access to systems.

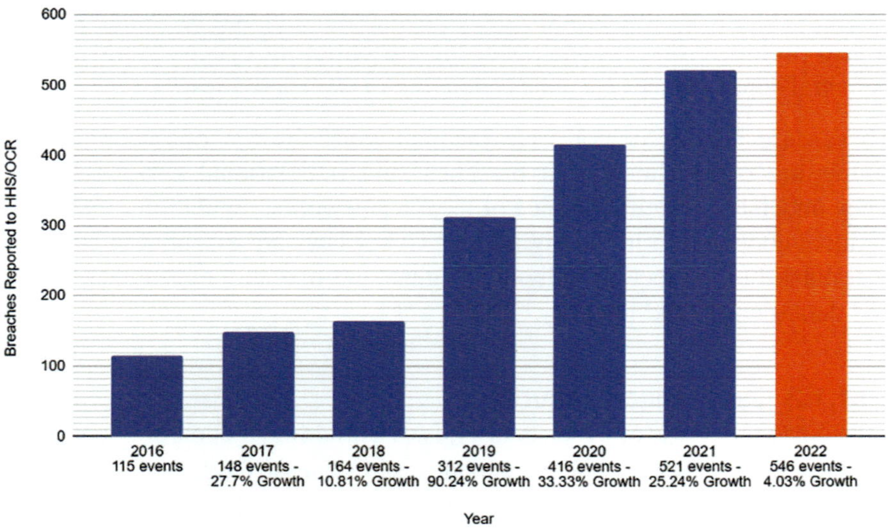

Figure 10.1 The cyber event in 2016–2022.

Data breaches: The theft or exposure of sensitive information, such as personal data, financial records, and trade secrets, can result in significant harm to individuals and organizations. Breaches can lead to identity theft, financial fraud, or reputational damage.

Malware: Malicious software like viruses, worms, Trojans, and spyware can infect systems and steal data, disrupt operations, or serve as a vehicle for other cyber-attacks.

Ransomware: Cybercriminals use ransomware to encrypt a victim's data and demand a ransom for the decryption key. Paying the ransom is discouraged, as it does not guarantee data recovery and encourages further criminal activity.

DDoS attacks: Distributed denial-of-service attacks flood a target system or network with traffic, rendering it unavailable to legitimate users. These attacks can disrupt online services, causing financial losses and damaging an organization's reputation.

Insider threats: Malicious actions or negligence from employees or contractors can pose a significant risk to organizations. Insiders may intentionally steal data, sabotage systems, or unintentionally cause security incidents.

State-sponsored attacks: Governments and state actors engage in cyber-espionage and cyberwarfare to gain strategic advantage or disrupt the

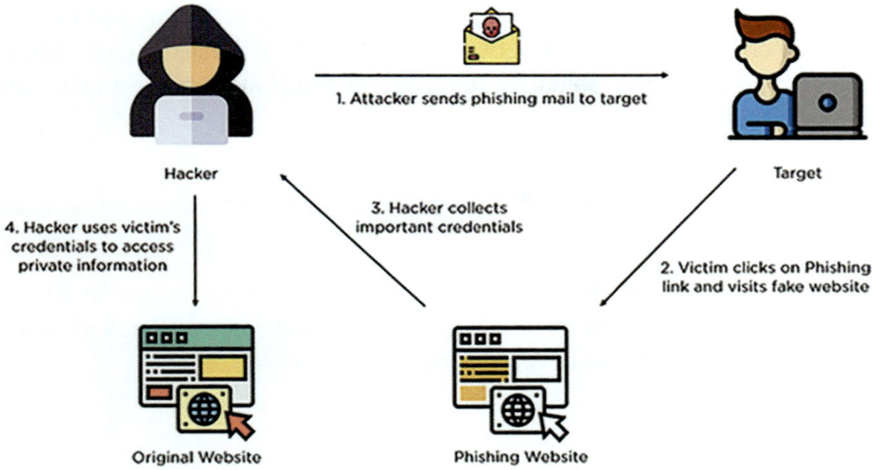

Figure 10.2 The phishing attack.

operations of other nations. These attacks often involve advanced and sophisticated techniques.

Phishing and social engineering: Phishing attacks use deceptive emails or websites to trick individuals into revealing sensitive information, such as login credentials or financial details.

Figure 10.2 shows the architecture of the phishing attack. Social engineering manipulates human psychology to gain unauthorized access to systems or data.

Internet of Things (IoT) vulnerabilities: The growing number of IoT devices often lacks robust security, making them attractive targets for cybercriminals. Compromised IoT devices can be used to launch large-scale attacks.

Supply chain attacks: Attackers may compromise software or hardware at various supply chain stages, introducing vulnerabilities or malware that can impact numerous organizations downstream.

Critical infrastructure vulnerabilities: Critical infrastructure, such as power grids, water supply systems, and transportation networks, increasingly relies on interconnected digital systems. Vulnerabilities in these systems can have serious real-world consequences.

Privacy concerns: The collection and misuse of personal data by corporations and governments are significant concerns. Privacy breaches and data abuse can result in identity theft, surveillance, and the erosion of personal freedoms.

Emerging technologies: As new technologies like artificial intelligence, quantum computing, and 5G networks advance, new threats and challenges emerge, including the potential for more sophisticated cyber-attacks and vulnerabilities.

International norms and cybersecurity: The need for universally accepted norms and rules for cyberspace can complicate international cooperation and responses to cyber threats.

Addressing these challenges and threats requires a multi-faceted approach, including solid cybersecurity practices, international cooperation, legislation, and ongoing research and development in cybersecurity.

10.1.2 Outer space

Like cyberspace, outer space presents significant security challenges due to the possibility of militarization and weaponization, which can pose defense and national security threats to all countries. International law governs outer space, considered a shared resource and cannot be claimed by any individual or nation. The 1967 Outer Space Treaty, ratified by nearly all countries engaged in space activities, establishes that the Moon and other celestial bodies should be utilized exclusively for peaceful purposes. It explicitly prohibits the deployment of weapons of mass destruction in space and the militarization of celestial bodies. The treaty places the onus on countries to promote responsible conduct in space and mitigate the likelihood of conflicts, even if private entities carry out such actions. However, there have been debates regarding interpreting the "peaceful use of space," especially as military activities, such as those observed during the Gulf War, have been conducted utilizing space-based assets. Satellite technology has become an integral part of military strategies for many nations. The complexity is further compounded by the emergence of space mining and tourism, as there needs to be a global consensus regarding the demarcation between airspace and outer space or the governing laws applicable in space.

10.1.2.1 Outer space and laws

Around 70–80 countries possess some level of sovereign space capability, allowing them to launch and operate their satellites independently. However, this also implies that approximately two-thirds of the world's nations need more national space capability and must rely on others to access space infrastructure and technology. The ability of these countries to utilize space technology for development and well-being is contingent on strategic and geopolitical relationships and agreements. Outer space security is becoming more important due to the increasing reliance on space-based technologies for various critical applications. The increasing dependence on space-based technologies for crucial applications has highlighted the vulnerabilities and challenges of safeguarding space infrastructure. Threats like anti-satellite weapons and space debris pose significant risks to national security and the global economy. International space law plays a vital role in addressing these challenges. The Outer Space Treaty, established in 1967, sets out fundamental principles for governing space activities. It prohibits the placement of weapons of mass destruction in outer space and celestial bodies and promotes the peaceful utilization of outer space. Efforts to prevent an arms race in space have been discussed by the Conference on Disarmament and the General Assembly since the 1980s. In 2017, a Group of Governmental Experts was formed to propose a legally binding instrument for preventing the placement of weapons in space, but a consensus could not be reached.

In 2020, the General Assembly adopted a resolution advocating for developing and implementing responsible behavior norms, rules, and principles in outer space to mitigate the risks of misunderstandings and miscalculations. Confidence-building measures (CBMs) in outer space have also been explored. In 1990, an expert group was tasked with studying CBMs, and their report was delivered in 1993. Since then, the General Assembly has initiated processes to develop transparency and CBMs in outer space activities. In 2010, governmental experts agreed on CBMs, including information exchange, risk reduction notifications, and international cooperation. Member states were encouraged to implement these measures through relevant national mechanisms. The UN-Space issued a special report in 2015 on the role of UN entities in supporting the implementation of CBMs, and the General Assembly requested the Secretary-General to submit a report on the coordination of CBMs in outer space activities, incorporating member states' views on these measures.

10.1.2.2 Recent developments in outer space exploration

Here, we list some recent programs identified in the developing stage executing in outer space exploration.

1. Mars exploration:
   - Perseverance rover: NASA's Perseverance rover successfully landed on Mars in February 2021. It is equipped with advanced scientific instruments and technology, including the Ingenuity helicopter, which achieved powered flight on Mars for the first time.

- Tianwen-1: China's Tianwen-1 mission also successfully reached Mars, deploying a rover named Zhurong in May 2021, marking China's first successful Mars rover mission.

2. Artemis program:
   - NASA's Artemis program aims to return humans to the Moon, with plans to land "the first woman and the next man" on the lunar surface by the mid-2020s. This program includes the Space Launch System (SLS) and the Orion spacecraft.

3. Private space industry:
   - Commercial space tourism: Companies like Blue Origin and Virgin Galactic have made significant progress in offering suborbital space tourism experiences to private individuals. SpaceX is also planning private missions, such as the Inspiration4 mission in 2021.
   - SpaceX's Starship: SpaceX has been developing its Starship spacecraft, which is intended to be a fully reusable spacecraft for missions to the Moon, Mars, and beyond. Several successful prototype test flights have been conducted [16].

4. International collaborations:
   - The International Space Station (ISS) continues to serve as a platform for international cooperation in space research and technology development. New international partnerships are being explored for future lunar and Mars missions [18].

5. Space telescopes:
   - The James Webb Space Telescope (JWST) is an upcoming space observatory expected to launch shortly. Observing the infrared spectrum will provide unprecedented insights into the universe.

Various space agencies and organizations are researching and planning for planetary defense missions to study and potentially deflect hazardous near-Earth objects (asteroids and comets).

These developments indicate a growing interest in outer space exploration, with government and private entities playing significant roles in advancing technology, scientific research, and space travel. Space exploration continues to be an exciting and evolving field with numerous opportunities and challenges.

10.1.2.3 Outer space security and challenges

Outer space security and the associated challenges have become increasingly significant as space activities have expanded. Following are some critical aspects of outer space security and the challenges they present.

Space debris: One of the primary challenges in space security is the proliferation of space debris, which includes defunct satellites, spent rocket stages, and fragments from previous collisions. This debris poses a collision risk to operational satellites and spacecraft, potentially causing damage and generating more debris.

Anti-satellite (ASAT) weapons: Various nations' development and testing of anti-satellite weapons pose a significant security challenge. These weapons have the potential to disable or destroy satellites in orbit, disrupting communication, navigation, and surveillance systems.

Cybersecurity: Space systems and satellite networks are vulnerable to cyber-attacks. These attacks can involve hacking into satellite communications, disabling satellite operations, or disrupting ground-based satellite control systems.

Jamming and spoofing: Jamming and spoofing interfere with satellite signals. Jamming involves disrupting communications by emitting radiofrequency interference, while spoofing involves sending false signals to deceive satellite receivers.

Secure satellite communications: The need for secure and encrypted satellite communications is growing, particularly for military and government users. Ensuring the confidentiality and integrity of sensitive data transmitted through space-based systems is challenging.

Space traffic management: With an increasing number of satellites and objects in space, ensuring safe and orderly operations is challenging. Space traffic management and collision avoidance are becoming crucial for maintaining space security.

Treaty compliance: The Outer Space Treaty, signed in 1967, prohibits the placement of nuclear weapons in space, among other provisions. Ensuring compliance with such treaties and norms is a challenge, especially with the growing militarization of space.

Orbital congestion: Limited orbital slots and frequencies are becoming congested, particularly in geostationary orbit. This congestion can lead to interference and conflicts between satellite operators.

Sustainability and long-term use: Ensuring the sustainability of space activities is essential. Proper disposal of satellites and upper stages and preventing on-orbit collisions are challenges in maintaining the long-term use of space.

Resource scarcity and conflicts: As space resources like minerals, water, and rare metals become targets for exploration, concerns about resource scarcity and potential conflicts over these resources arise.

Space traffic and collision avoidance: To prevent accidental collisions in crowded orbits, coordination and collision avoidance measures among satellite operators are crucial.

Space situational awareness: Developing accurate and comprehensive awareness is crucial for identifying potential threats, tracking space debris, and ensuring safe space operations.

National security: National security interests increasingly rely on space assets for communication, surveillance, navigation, and more. Protecting these assets from potential threats is a growing challenge.

Addressing these challenges and ensuring outer space security requires international cooperation, transparent policies, and the development of norms and rules for responsible space activities. The militarization of space also necessitates careful consideration of arms control agreements and efforts to prevent the weaponization of space.

10.2 Literature Review

10.2.1 Cyberspace

Cybersecurity is a subset of IT security concerned with protecting digital data from unauthorized access, attacks, and destruction. Brenner [1] proposed a classification of three main categories of cyber harms: individual, systemic, and apprehension. Kshetri [2] developed a cost–benefit analysis considering the characteristics of cybercriminals, victims, and law enforcement officials. [3] The crime triangle is a framework that defines cybercrime by requiring a victim, motive, and opportunity. [4] Machine learning and information-digging approaches can assist in detecting interruptions during digital investigations.

While cyber-attacks have become more targeted and sophisticated, frequent untargeted attacks still exploit vulnerabilities. [5] "Opportunistic attacks" refer to attacks that take advantage of these vulnerabilities. A new encryption method called Camellia has been introduced, which provides robust security and efficiency for software and hardware platforms. Camellia's encryption speed is comparable to other AES finalists. In a study in [6], sentiment analysis and machine learning techniques were used to identify cyber risks that were previously challenging to detect. Greenfield and colleagues [7] proposed a methodology for evaluating harm, considering dimensions such as functional integrity, material support, and freedom from humiliation, privacy, autonomy, and reputation. Grant et al. [8] introduced the concept of "cyberspace cartography" for military operations, and Mathieu and Guy emphasized the importance of high-quality literature reviews. Grant et al. coined the term "cyberspace cartography" and applied the concept of "cyber-geography" to military operations. They also suggest that their ontology might be used in research to help solve the attribution problem of being unable to quickly identify hostile actors in cyberspace. In [9], the issue of Internet jurisdiction law and determining legal authority in multi-state lawsuits are discussed. Four potential approaches to determining controlling jurisdiction are presented based on factors such as the citizenship of the offending data or system's subject, the location of harm, or the citizenship of the data creator or holder. Ensuring up-to-date intelligence on cybersecurity risks and vulnerabilities is crucial for protecting personal and organizational systems. Efforts are ongoing to improve risk frameworks and model the resilience of business systems. Updating security software, providing simulation-based training scenarios, and continuous and timely software updates are suggested. Various studies have utilized machine learning and data mining techniques for cyber analytics and intrusion detection. Edwards et al. [10] use a publicly available dataset of data breaches to uncover trends in data breaches using a Bayesian generalized linear model. They conclude that while the amount and frequency of data breaches have remained consistent in recent years, their impact is increasing as threat actors improve their ability to monetize personal information and the quantity of electronic financial transactions grows. A concentrated literature analysis of machine learning and data mining methods for cyber analytics in support of intrusion detection was reported in a survey study [11]. A "top-down" approach in criminology was employed to elicit premiums that users would pay to protect their assets from cyber incidents. In the paper [12], the author used a semi-supervised method to classify cybersecurity logs into three categories: attack, unsure, and no attack, by first breaking the data into three clusters using fuzzy K means (FKM), then manually labeling a small sample, and finally training the neural network classifier multi-layer perception (MLP) on the manually labeled data. An interesting approach, based on the "top-down" methodology described in the criminology field, is presented by Nguyen et al. [13]. Cyber-attacks threaten patient safety and the healthcare industry due to compromised medical data integrity and disruption of medical devices. Social media data, sentiment analysis, and machine learning are used to analyze cyber threats and opinions. In [14], the healthcare industry's vulnerability to cyber threats is emphasized, and the use of machine learning to detect cyber threats is investigated. Cyber-attacks are prevalent in various domains, including the travel industry and cryptocurrency exchanges. The global cost of cybercrime is estimated to be significant, and awareness campaigns are essential in preventing or reducing cybercrime.

10.2.2 Outer space

Space exploration began in 1957 when the Soviet Union launched Sputnik, the first artificial satellite to orbit Earth. This occurred during the Cold War, a period of political tension between the Soviet Union and the United States, where both nations were striving to develop intercontinental ballistic missiles (ICBMs) capable of delivering nuclear weapons across continents. Soviet rocket engineer Sergei Korolev played a crucial role in developing the R7 rocket, the first intercontinental ballistic missile (ICBM) that marked the start of the space race. The launch of the satellite Sputnik by the Soviets, carried by an R7 rocket, created concerns for the United States as it demonstrated the advanced technological capabilities of the Soviet Union. The Soviet Union achieved further milestones by sending the first human, Yuri Gagarin, into space in 1961. This prompted President John F. Kennedy to challenge the United States to land a man on the moon, leading to the successful Apollo 11 mission in 1969. During the 1960s and 1970s, the United States and the Soviet Union made significant progress in space exploration. They launched probes to study Venus, Mars, and Mercury, established space stations, and made discoveries about moons around Jupiter and Saturn potentially harboring life. However, this period also witnessed the beginning of space militarization, with the deployment of anti-satellite weapons.

In recent years, China, the United States, and India have demonstrated their capabilities to weaponize space through various methods, such as kinetic anti-satellite weapons [15], lasers, cyber capabilities, and electromagnetic pulse weapons. These developments have raised concerns about the potential for an arms race in outer space and the use of destructive space weapons. Existing international agreements, including the Outer Space Treaty, need to be revised to address the challenges posed by space weaponization [17]. The treaty prohibits the testing and development of weapons of mass destruction in space but does not place restrictions on other kinetic and non-kinetic weapons. Proposed treaty amendments have faced opposition and have yet to

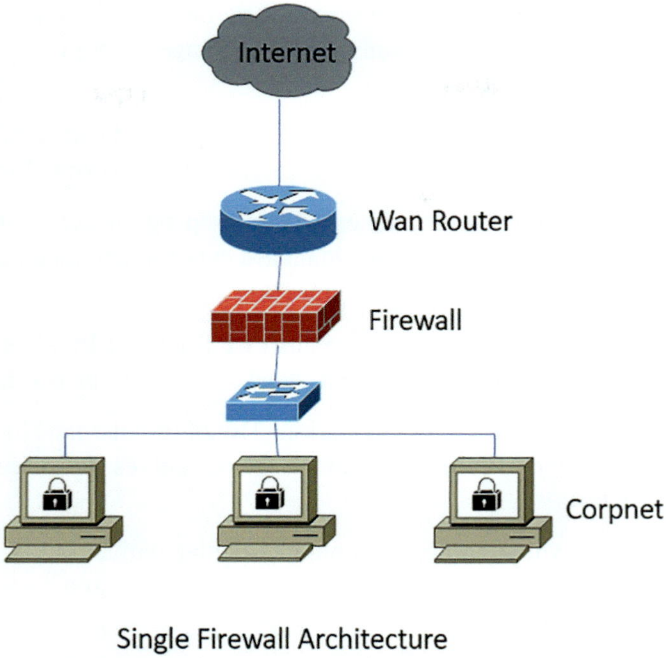

Figure 10.3 A simple firewall architecture.

address the dual-use nature of space technologies and ground-based attacks on space assets.

To address these issues and prevent further risks, there is a need for a comprehensive arms control agreement that bans the use of weapons against space objects and the use of space objects as weapons. This agreement should also prohibit space weapons development, testing, and deployment. Updating international law to reflect the current realities of space exploration and weaponization is essential to ensure outer space's peaceful and responsible use.

10.3 Methodologies

Different methodologies to prevent outer space attacks: Firewalls serve as a protective barrier between internal networks and the Internet, controlling network traffic based on predefined rules. A simple firewall architecture is shown in Figure 10.3.

Ensuring access control: Proper access control methods, like strong passwords, two-factor authentication, and role-based access, help prevent unauthorized entry to sensitive information or systems.

- Utilizing encryption: Encryption converts data into a coded language, safeguarding it during transmission and storage by requiring a specific key for decryption.
- Managing software patches: Keeping software updated with the latest security patches helps prevent the exploitation of known vulnerabilities.
- Employee training and awareness: Training programs educate employees on identifying and avoiding standard cyber-attack techniques, such as phishing scams and social engineering.
- Establishing incident response plans: Well-defined incident response plans mitigate the impact of cyber-attacks and limit further damage.
- Implementing network segmentation: Dividing networks into smaller isolated segments limits cyber-attack spread and restricts access to critical systems.
- Conducting vulnerability scanning: Regular vulnerability scanning identifies potential security weaknesses, enabling proactive measures to address them.
- Performing penetration testing: Simulating attacks on systems or networks through penetration testing helps identify and address security gaps before they can be exploited.
- Deploying security tools: Utilizing security tools like antivirus software, intrusion detection systems, and web application firewalls aids in preventing and detecting cyber-attacks.

10.4 Applications

Cyberspace and outer space have various applications critical to national security, commerce, and everyday life. Here are some of the main applications of cyberspace and outer space:

Applications of cyberspace:

- Communications: The Internet and other cyberspace networks enable fast, efficient communication between individuals, businesses, and governments.
- Commerce: Cyberspace has become a critical commerce component, enabling online shopping, banking, and other transactions.
- Critical infrastructure: Many critical infrastructure systems, such as power grids and transportation networks, rely on computer systems and networks.

- National security: Cyberspace is a critical domain in national security, with military and intelligence agencies using cyberspace to collect intelligence, carry out cyber-attacks, and defend against cyber threats.
- Education: Online learning and training have become common uses of cyberspace, enabling people to access educational resources anywhere in the world.

Applications of outer space:

- Communications: Satellites provide a critical role in global communications, enabling the transmission of data and voice communications across the globe.
- Earth observation: Satellites provide vital data on weather patterns, natural disasters, and other environmental conditions, which can be used for scientific research, disaster management, and agriculture.
- Navigation: Satellites provide the backbone for the GPS, which enables accurate navigation and timing for military and civilian applications.
- National security: Space-based assets are also critical to national security, providing capabilities such as early warning of missile launches and satellite-based reconnaissance.
- Scientific research: Space-based telescopes and other scientific instruments are used for a wide range of scientific research, such as studying the universe's origins, searching for exoplanets, and studying Earth's climate.

10.5 Cyberspace Security Measures Within the Domain of Outer Space

Cybersecurity is crucial in various domains, and outer space is no exception. As our reliance on space-based systems and technologies increases, ensuring the security of these systems becomes essential. Here are some ways in which cybersecurity can be used in outer space:

1. Secure communication: Cybersecurity measures can be implemented to protect the communication channels used in outer space. Encryption techniques can be applied to secure data transmissions between satellites, space probes, and ground stations, preventing unauthorized access or tampering.
2. Protection against cyber-attacks: As space-based systems become more interconnected and reliant on software, they become potential targets

for cyber-attacks. Implementing robust cybersecurity measures can help protect these systems from unauthorized access, data breaches, or malicious activities. This includes techniques like firewalls, intrusion detection systems, and regular security audits.

3. Satellite and spacecraft security: Satellites and spacecraft can be vulnerable to cyber threats. Configures can be minimized by implementing secure coding practices and conducting rigorous security testing during development. Additionally, authentication protocols and access controls can be employed to prevent unauthorized access to critical systems.
4. Mission data protection: Outer space missions generate vast amounts of sensitive data. Cybersecurity measures can safeguard this data from interception, manipulation, or theft. Data encryption, secure storage systems, and secure data transfer protocols can be used to protect mission-critical information.
5. Ground station security: Ground stations are vital for communicating with and controlling satellites and spacecraft. Strong cybersecurity measures at ground stations are crucial to prevent unauthorized access or interference. This includes secure network infrastructure, access controls, and monitoring systems to detect and respond to potential threats.
6. Space traffic management: With the increasing number of satellites and space objects in orbit, space traffic management becomes crucial to avoid collisions and ensure the long-term sustainability of space activities. Cybersecurity measures can help protect space traffic management systems from cyber-attacks, ensuring the integrity and reliability of the information used for orbital predictions and collision avoidance.
7. Cybersecurity awareness and training: Promoting cybersecurity awareness among space agencies, astronauts, and mission personnel is essential. Training programs can educate individuals about best practices for secure operations, recognizing potential threats, and responding effectively to cyber incidents.

Overall, integrating cybersecurity into outer space activities is vital to protect critical space-based systems, ensure the confidentiality and integrity of mission data, and mitigate the risks associated with cyber threats in the space domain.

Table 10.1 categorizes the difference between different aspects that are looked upon for security in both domains. This table highlights critical distinctions between cybersecurity and outer space security, underlining their

Table 10.1 Cybersecurity vs. outer space security challenges.

Aspect	Cybersecurity	Outer space security
Focus	Protecting digital systems, networks, data	Safeguarding assets, operations in space
Scope	Computers, networks, data, IoT devices	Satellites, spacecraft, space stations
Threats	Hacking, malware, data breaches	Space debris, collisions, interference
Importance	Safeguarding digital information	Ensuring space assets' safety
Applications	Data protection, network security	Satellite protection, space exploration
Challenges	Evolving cyber threats, system vulnerabilities	Proliferation of space debris, lack of transparent laws
International Aspect	Global concern, international cyber laws	Need for international space regulations
Impact	Financial loss, data compromise	Loss of critical communication, navigation
Prevention	Firewalls, encryption, regular updates	Space traffic management, debris removal

differing focuses, challenges, and approaches to ensuring safety and integrity within their respective domains.

10.6 Conclusion

In conclusion, the intertwining realms of cyberspace and outer space necessitate a comprehensive approach to security. The increasing reliance on space-based systems and technologies underscores the importance of implementing robust cybersecurity measures to safeguard critical infrastructure and operations. To effectively address security concerns, it is essential to establish secure communication channels within outer space. Encryption techniques should be employed to protect data transmissions between satellites, space probes, and ground stations, minimizing the risk of unauthorized access or tampering. Defending against cyber threats is paramount in safeguarding space-based systems. This involves implementing advanced firewalls, intrusion detection systems, and security audits to prevent unauthorized access, data breaches, or malicious activities. By prioritizing secure coding practices and rigorous security testing during the development of satellites and spacecraft, vulnerabilities can be minimized, ensuring their resilience against potential cyber-attacks. The protection of mission data is of utmost importance in outer space operations. Implementing data encryption, secure storage systems, and robust data transfer protocols safeguards sensitive information from interception, manipulation, or theft, thereby preserving the integrity and confidentiality of mission-critical data. Ground station security is also crucial in ensuring the overall security of outer space activities. Robust cybersecurity measures should be applied to ground stations, including secure network infrastructure, access controls, and monitoring systems. The reliability and integrity of communication with satellites and spacecraft can be maintained by preventing unauthorized access and interference.

Furthermore, effective space traffic management is vital to avoid collisions and ensure the long-term sustainability of space activities. Cybersecurity measures play a significant role in protecting space traffic management systems from cyber-attacks. The safety and efficiency of space operations can be maximized by maintaining the integrity and reliability of information used for orbital predictions and collision avoidance. Finally, promoting cybersecurity awareness and providing training to space agencies, astronauts, and mission personnel is crucial. Educating individuals about best practices for secure operations, raising awareness about potential threats, and fostering the ability to respond effectively to cyber incidents strengthen the overall cybersecurity posture in outer space. In conclusion, integrating cyberspace security measures in outer space is essential to protect critical space-based systems, preserve the integrity of mission data, and mitigate the risks posed by cyber threats. By embracing comprehensive security practices, we can confidently navigate the boundless frontiers of cyberspace and outer space, enabling continued exploration and utilization while ensuring a safe and sustainable environment beyond Earth's atmosphere.

Acknowledgments

I want to express my sincere appreciation to Dr P. Karthikeyan for his continuous guidance and support in preparing chapters of the book.

I also thank Dr . C. S. Ramesh, Presidency University, for his insightful comments and valuable feedback, which significantly improved the quality of this chapter.

Lastly, I would like to acknowledge the support and understanding of my family and friends who stood by me while completing this work.

References

[1] Brenner, S.W. Cybercrime metrics: old wine, new bottles? Va. JL & Tech, 9:13–13, 2004.

[2] Kshetri, N. The simple economics of cybercrimes, IEEE Secur Priv, 4, pp. 33–39, 2006.

[3] Dhanjani, N., Rios, B., and Hardin, B. Hacking: The Next Generation: The Next Generation. O‟Reilly Media, Inc., 2009.

[4] Cross M., and Shinder, D.L. Scene of the cybercrime.Syngress Pub., 2008.

[5] Perwej, Y., Haq, K., Jaleel, U., and Parwej, F. "Block ciphering in KSA, A major breakthrough in cryptography analysis in wireless networks", International Transactions in Mathematical Sciences and Computer, India, ISSN-0974-5068, Volume 2, No. 2, Pages 369–385, July-December 2009.

[6] Fink, E., Sharifi, M., and Carbonell, J.G. "Application of machine learning and crowdsourcing to detection of cybersecurity threats", In Proceedings of the US Department of Homeland Security Science Conference–Fifth Annual University Network Summit, Washington, DC., 2011.

[7] Greenfield, V.A., and Pa. L. A framework to assess the harm of crim. Br J Crimi., vol. 53, pp. 864–885, 2013.

[8] Grant, T., and Liles, S. On the military geography of cyberspace," Proc. Int. Conf. Inf.Warfa, p. 66, 2014.

[9] Chertoff, M., and Rosenzweig, P. (Mar. 1, 2015). A Primer on Globally Harmonizing Internet Jurisdiction and Regulations, accessed on oct. 15, 2015.

[10] Edwards, B., Hofmeyr, S., and Forrest, S. Hype and heavy tails: a closer look at data breaches. J Cyber secur 2016;2:3–14.

[11] Buczak, A.L., and Guven, E. "A survey of data mining and machine learning methods for cyber security intrusion detection. IEEE Communications Surveys & Tutorials, 18(2), pp. 1153–1176, 2016.

[12] Teoh, T.T., Zhang, Y., Nguwi, Y.Y., Elovici, Y., and Ng, W.L. "Analyst intuition inspired high velocity big data analysis using PCA ranked fuzzy k-means clustering with multi-layer perception (MLP) to obviate cyber security risk ", 13th International Conference onNatural Computation, Fuzzy Systems and Knowledge Discovery (ICNC-FSKD), pp. 1790–1793, IEEE, 2017.

[13] Nguyen, K.D., Rosoff, H., and Richard, S.J. Valuing information security from a phishing attack.In: International Conference on Applied Human Factors and Ergonomics. Cham: Springer, 2017.

[14] Furnell, S., and Emm. D. "The ABC of ransomware protection", Comp. Fraud & Sec. (10), pp. 5–11, 2017.

[15] Ying-Yu Lin. "China Cyber Warfare and Cyber Force." Tamkang Journal of InternationalmAffairs , vol. 22, no. 3, pp. 119–161, 2019.

[16] Amos, J. (2019). SpaceX Puts Up 60 Satellites. BBC News. Retrieved September26,2019,from:https://www.bbc.com/news/science-environment-48289204.

[17] Andersson, J., and Malm, A. (2006). Public–Private Partnerships and the Challenge of Critical Infrastructure Protection. In M. Dunn and V. Mauer (eds), International CIIP Handbook 2006, Vol. II: Analyzing Issues, Challenges, and Prospects. Zurich: Swiss Federal Institute of Technology, pp. 139–167.

[18] Araxia Abrahamian, A. (2017, September 15). How a Tax Haven is Leading the Race to Privatise Space. The Guardian. Retrieved September 26, 2019, from: https://www.theguardian.com/news/2017/sep/15/luxembourg-tax-haven-privatise-space.

Index

About the Editors

Dr. P. Karthikeyan obtained his Bachelor of Engineering (B.Eng.,) in Computer Science and Engineering from Anna University, Chennai, and Tamil Nadu, India in 2005 and received his Master of Engineering (M.Eng.) in Computer Science and Engineering from Anna University, Coimbatore India in 2009. He completed his Ph.D. degree in Anna University, Chennai in 2018. He works as post doctorate research fellow at the department of computer science and information engineering, National Chung Cheng University, Taiwan, and contributed to the national projects on using technology to promote human rights and sustainable development, supported by the National Science and Technology Council, Taiwan. He possesses a high level of proficiency in project development and research, specializing in the domains of Cloud computing and the practical application of deep learning. He is well-versed in programming languages, including Java, Python, R, and C. His research endeavours have led to the publication of more than 30 papers in esteemed international journals, many of which have garnered commendable impact factors. Furthermore, he has shared his research insights through presentations at over 20 international conferences, solidifying his reputation within the academic community.

Polinpapilinho F. Katina is an Assistant Professor in the Department of Informatics and Engineering Systems at the University of South Carolina Upstate (Spartanburg, South Carolina, USA). He previously served as a Postdoctoral Researcher for the National Centers for System of Systems Engineering (Norfolk, Virginia, USA) and Adjunct Professor in the Department of Engineering Management and Systems Engineering at Old Dominion University (Norfolk, Virginia, USA). Dr. Katina holds a B.Sc. in Engineering Technology with a minor in Engineering Management, an M.Eng. in Systems Engineering, and a Ph.D. in Engineering Management/ Systems Engineering, all from Old Dominion University (Norfolk, Virginia, USA). He has received additional training from, Politecnico di Milano (Milan, Italy). His areas of research/teaching interests include, among others, complex system governance, critical infrastructure systems, decision making and analysis (under uncertainty), emerging technologies, energy systems (smart grids), engineering management, infranomics, manufacturing systems, system-of-systems, systems engineering, systems pathology, and systems thinking.

Rajagopal Ramasamy is currently working as an associate professor in the Department of Computer Science and Engineering at Alliance University, Bengaluru. He completed his Ph.D. degree at Anna University, Chennai. He received his post-graduation and bachelor's degrees in computer science and engineering from Anna University, Chennai. He has served as a reviewer for Inder Science and other prestigious Scopus-indexed journals. He serves on the editorial board of the International Journal of Mathematical Sciences and Engineering. He is a life member of the Indian Society for Technical Education (ISTE). He has published in more than 50 international journals and presented at many international conferences. His research interests include cloud computing, high-performance computing, big data analytics, and algorithms.

For Product Safety Concerns and Information please contact our EU representative GPSR@taylorandfrancis.com Taylor & Francis Verlag GmbH, Kaufingerstraße 24, 80331 München, Germany